Material Culture and Electronic Sound

Material Culture and Electronic Sound

artefacts
STUDIES IN THE HISTORY OF SCIENCE AND TECHNOLOGY
VOLUME 8

EDITED BY

Frode Weium and Tim Boon

MANAGING EDITOR

Martin Collins, Smithsonian Institution

Smithsonian Institution
Scholarly Press

WASHINGTON, D.C.

2016

Published by
SMITHSONIAN INSTITUTION SCHOLARLY PRESS
P.O. Box 37012, MRC 957
Washington, D.C. 20013-7012
www.scholarlypress.si.edu

Front cover image: Charles Parsons's Auxetophone-Gramophone (ca. 1905). Courtesy of the Science Museum.

Back cover images: Left: Varispeed Revox A77 tape recorder used by electroacoustic musician Hugh Davies. Courtesy of the Science Museum. Right: The Oramics Machine tone generator cabinet. Courtesy of the Science Museum.

Library of Congress Cataloging-in-Publication Data

Material culture and electronic sound / edited by Frode Weium and Tim Boon.
p. cm. — (Studies in the history of science and technology ; v. 8)
Includes bibliographical references and index.
1. Electronic music—History and criticism. 2. Electronic musical instruments—History.
3. Music and technology. I. Weium, Frode. II. Boon, Tim.
ML1092.M37 2013
786.7—dc23 2012032373

ISBN: 978-1-944466-08-4 (pbk)
ISBN: 978-1-935623-10-6 (cloth)

Printed in the United States of America

∞™ The paper used in this publication meets the minimum requirements of the American National Standard for Permanence of Paper for Printed Library Materials Z39.48–1992.

Contents

Series Preface

SCIENCE AND technology have been a defining element of the modern era. They have shaped our lives in large and small ways—through broad understandings of the universe and in the tools and objects that make up the texture of everyday life. They have been preeminent activities for organizing expertise and specialized knowledge; in defining power and progress; and in shaping the development of nations and our relations with others across the planet. In 1996, the Smithsonian, the London Science Museum, and the Deutsches Museum formed Artefacts to emphasize the distinctive role that museums through their collection, their display, and, especially, their study of objects can play in understanding this rich and significant history.

Artefacts has two primary aims: to take seriously the material aspects of science and technology through understanding the creation and use of objects historically, and to link this research agenda to the exhibition and educational activities of the world community of museums concerned with the intimate connections among material culture, the history of science and technology, and the transnational. The effort gradually has gained footing: Artefacts holds an annual conference and has expanded its formal organization to include fourteen cosponsors (listed on this volume's copyright page). This expanded community, composed primarily of European and North American museums, provides opportunity for more robust professional conversation and, not least, broadens the range of local and national historical experiences of science and technology represented in Artefacts. Not least, Artefacts has created a fruitful interplay between scholarly research and museum practice: Aided by its Advisory Editorial Board, it publishes this book series, which, in conjunction with annual meetings, has helped stimulate a broader turn toward material-based research in scholarship and its use in museum collecting and exhibitions.

The Artefacts community believes that historical objects of science and technology can and should play a major role in helping the public understand science and technology—the ingenuity associated with these activities, their conceptual underpinnings, their social roles, their local and global connotations. We welcome other museums and academic partners to join our effort.

Martin Collins
National Air and Space Museum, Smithsonian Institution
Series Managing Editor

Foreword

Brian Eno

MUSIC AFTER electricity is to music as cinema is to theater. Most technical innovations in music are made for historical reasons: something is developed to solve an existing problem, to do an existing job cheaper or faster or more portably or better. Instrumental amplification, for example, came into being to allow some rather quiet instruments such as guitars and basses to be heard in large ensembles. Multitrack recording appeared so that engineers didn't have to fix the balance between, say, a vocalist and a band during the recording session itself, but could, by recording them on separate tracks, defer the decision until later. What inevitably happens, however, with art and technology is that an invention made for historical reasons then turns out to suggest new possibilities that were not even thought about prior to the technology existing. Technology suggests new places for art to go. Amplification, it turned out, didn't only make instruments louder: it also gave them access to a huge tonal palette that had previously never been suspected—amplification created new instruments. Multitracking in turn made it possible to develop a composition over a period of time, rather than having to commit to a simultaneous performance of it. Music moved from being an ephemeral medium caught in a moment in time and became more like painting—a process extended over noncontiguous time.

These are just two of the myriad ways that electricity changed music. In fact, the sum of these changes—which are still unfolding—have made a medium so different that we really should have come up with another name. Just as we don't call cinema *theater*—despite their common roots—we shouldn't really be calling the new medium of electronically created sound *music*. It shares very few characteristics with its ancestor: This music is different in every sense except for the fact that it enters your sensorium via your ears. It is made differently, from different materials in different ways, by different people, to be heard in different places in different ways and at different times. It is arguably further from its "live" ancestor than cinema is from theater.

Perhaps the revolution of electronic music could be summed up as the *materialization* of sound. What happened over the last century or so is that it became possible for artists to work with sound as a plastic, malleable material. Recording onto tape took sound out of time and brought it into space—made it a substance that could be stretched, reversed, edited, and worked on in numerous other ways. Because of amplification, the revolution was also about scale. Acoustic instruments, to be audible at all, require the player to move large quantities of molecules. This means that those instruments are designed from the group of physical possibilities that could produce loudness. Amplification makes it possible to design instruments at the atomic level: Instead of moving a great lump of physical material, you now only have to move a few atoms and amplify them—an electric guitar is really just a moving string, disembodied . . . and a synthesizer is a way of creating minute atomic changes within a few transistors. And we are already beyond these stages, beyond molecules and atoms and into data. Modern instruments—plug-in synthesizers, for example—are ways of moving information about.

It isn't until the electrical impulse hits the loudspeaker cone and causes it to begin moving air that "sound" is produced. "Real" instruments, even the amplified ones, start with some physical event that makes sound. Synthesizers don't: they start with processes that will produce electrical signals that can drive loudspeakers. So modern electronic instruments are incomplete without their "bodies"—the amplifiers and loudspeakers by which they're heard. Those bodies include the amps and speakers in a listener's home: they're as much a part of the instrument as any other, since it is only through them that they are audible. Before the amp and speakers, it's all just numbers.

Modern electronic music isn't "figurative": it doesn't have to call to mind some pre-existing reality (e.g., "grand piano" or "woodwind"), but instead can create new—and physically impossible—ones. As a medium, then, it's similar to animated film: instead of recordings now being documentations of real instruments played in real time, they are instead increasingly frame-by-frame constructions with sound, rather than image, as the material. Perhaps a better name for this new medium would be *animated sound.*

When Steinway introduced a third pedal on their grand pianos, it suggested a new way of composing, which was taken up with great enthusiasm by Debussy, among others. Revolutions of similar scale now happen almost weekly in the ever-accelerating evolution of contemporary composition. There's a lot to look forward to!

Introduction

Tim Boon and Frode Weium

WE INHABIT a sound world radically distinct from that of our forebears. The instruments and devices we use to originate, record, and reproduce music have changed what we hear and how we hear it, to the extent that what counts as music now effectively knows no sonic bounds. Both the forms of music and where we can expect to hear it have been transformed. This is a fundamental revolution that can be traced to the specific forms of musical instruments and audio devices. The Artefacts series is devoted to research into the material culture of science, technology, and medicine. Music is an especially potent subject for the series because the cultural product—the music we hear—has been so dramatically affected by technology. Accordingly, the chapters that follow have a particular focus on the instruments and devices themselves; using examples from museum collections, the authors investigate questions about the relationship between the specific forms of musical instruments—and other devices implicated in its production and distribution—and the musical possibilities actualized by performers. The chapters also demonstrate how close examination of objects and detailed attention to museum collections can pose new questions and encourage new perspectives.

Musicology and the history of technology have often been perceived as separate fields. However, there has been an increased interest within both disciplines in studying the relationship between music and technology, and in particular the relationship between music and the development of electronic musical instruments and devices. In addition to some broader works,[1] a number of monographs has also been published.[2] To varying degrees, these works have been preoccupied with objects as material expressions of human culture. But there is still a real need for studies that go beyond strict technical descriptions on the one hand and an overly broad social history on the other. Here we aim to make a distinctive museological contribution, joining Trevor Pinch and Karin

Bijsterveld on the new territory of "Sound Studies," which they eloquently laid out in 2004.[3] The chapters in this volume focus on the musical instruments and devices themselves (their origin, mode of operation, use, reception, and so on), so as to communicate new insights about the relationship between music and technology, and also to map social and cultural change in a broader perspective.

Much of the revolution that has occurred in music since the turn of the twentieth century has been related to the application of electricity and electronics. This volume's major concern is the extent to which this marks a distinct phase in the development of music, or whether the application of these new technologies has merely exaggerated existing tendencies or simply multiplied musical choices for both performers and listeners. The chapters in this volume show that in music, as in all areas of culture, there is continuity as well as change, and electronics may not have been the first means to a particular sonic effect. In particular, Aleksander Kolkowski and Alison Rabinovici in their detailed exploration of Parsons and Short's Auxetophone show not only how amplification antedated the electronic valve but also how its application prefigured some later debates. In a counterfactual world, would Bob Dylan's adoption of compressed air amplification have been as scandalous as his "going electric"? Quite possibly. The power of this chapter is to show that those intent on illuminating the interaction of culture and technology should be wary of assigning simple cause and effect. Equally, the interactions between old and new technologies provide rich ground for new work that illuminates these same relations. Katy Price demonstrates how artifacts used in performances can enrich audience awareness of the history of technology. By analyzing the use of Stroh instruments, a pianola, and an iPhone in two different performances, she also illuminates whether the application of electronics does mark a distinct phase in the development of music. In counterpoint to Brian Eno's point in our foreword, where he argues that electricity has changed music "in every sense except for the fact that it enters your sensorium via your ears," Price's chapter suggests that the transition has not been linear, but rather should be understood as complex and multifactorial.

The slow revolution that has transformed music undoubtedly seems profound to us because we are living through its cultural consequences. This is clear with wholly electronic musical genres such as techno and electropop. Other kinds of music have been stamped more tangentially, but nonetheless significantly; it is impossible to imagine dub reggae without the use of tape echo, for example. But this electronic revolution is not the only transformation in the history of music to be linked to the emergence of new instrumentation; both the efflorescence of the Baroque period and the stabilization of the symphony orchestra, for example, qualify to be considered similarly revolutionary.

For all that instrumental and musical novelty is much older than the electronic valve and transistor, there is scarcely a type of music that has not been affected by electronics, including the ostensibly least electronic of musical forms, the "classical" concert—even where amplification is not used, its audience reach is enormously extended by broadcasts that today rely on microphones, microelectronics, and analog-to-digital conversion. Recording allows for repeated listening and thereby enables new kinds of discernment. Technology may also suggest new compositional technique, as is clear in the example of contemporary classical composer Steve Reich, in which phase shifting commenced with tape pieces such as *It's Gonna Rain* (1965), was transferred to human performers in *Piano Phase* (1967) and on into the fabric of much of his later work, which for many years eschewed electronic means.[4] In general, there has been a technologization of music, including "classical music," or at least a shifting of its specific cultural form under the influence of technology.

An enduring theme in the history of electronic musical devices is the dialectic between imitation and novelty. This dialectic reached an unstable early resolution with the first electronic instruments, including the Theremin; despite their potential for creating new sounds, these were often used to imitate traditional instruments, including the cello. The same can even be said of today's computers/software.[5] The point of departure for Frode Weium's chapter is that new music technologies have always sparked debate and that they have often been rejected on account of their artificiality. His chapter examines the reception of the electromechanical Hammond organ in Norway from the mid-1930s up to the 1960s. Some of the first imported organs have been traced and are used to illustrate how the instrument was adapted to different environments, including churches. Weium discusses the notion of the instrument as an artificial surrogate for the pipe organ and shows how early enthusiasm turned to hostility. Other devices, such as string synthesizers, have been made and sold on the basis that they can replace the function of conventional musicians. Sarah Angliss's wide-ranging chapter shows how key new imitative musical technologies—notably drum machines and samplers—were received by musicians and organizations, provoking debates about authenticity and threatening to replay arguments about automation and unemployment. She shows how rich the interaction of musicians and machines became as some musicians came to favor nonimitative aspects of these devices—at one extreme, the robotic character of drum machines, and at the other, types of complexity that machines offered that humans could not easily replicate.

By contrast with imitative devices, there is the creation of sounds that have never been heard before, whether that has been achieved with great difficulty by tape splicing and varispeed playback, as practiced in the early days of the BBC Radiophonic Workshop, or with comparative ease using voltage-controlled synthesizers.[6] The magnetic tape recorder had a profound impact on music in the twentieth century. While some used it in similar ways as earlier recording devices, others took advantage of the changes it enabled. Ragnhild Brøvig-Hanssen argues that it represents an important shift to what she, following R. Murray Schafer, calls a new era of "schizophonia," where multitracking extended the ability to manipulate the space and time component of recorded music. Using examples including a tape recorder and loops from the Hugh Davies collection at the Science Museum in London, she identifies alternative recording paradigms. Here again we find devices created for one purpose being pressed into new kinds of musical service. Sean Williams, in his close examination of the use of filters by two music makers as different as Karlheinz Stockhausen and King Tubby, documents how these technical devices—just as the magnetic tape recorder—were used as dynamic musical instruments, quite differently from their original intended purpose. Noises and environmental sounds emphasized the physicality of the performance and the material nature of music making. As Williams writes, "This kind of material research can provide a solid phenomenological foundation for further musicological, sociological, or anthropological studies."

Technology museum curators often worry about the problems of displaying electronic devices in a compelling way, compared with the "brass and glass" of older scientific instruments that convey their function more directly. One black box looks much like another, whether it is a DNA synthesizer or a rack-mounted music studio peripheral device. It may be that paying attention to musical devices is fertile territory for the exploration of these issues. The design of a Minimoog, for example, in the design of its circuits—voltage-controlled oscillators, filters, and amplifiers—embodies a set of propositions about the characteristics of natural sounds and how they may be artificially produced. When rendered virtually in "soft synthesizer" programs, the original analogies carry over. In other words, electronic musical devices are highly legible as embodied analogies. The Oramics Machine from the Science Museum's collection provides a special case of this argument, as Mick Grierson and Tim Boon show. Daphne Oram's unique device controlled compositional elements, pitches, and timbre using a unique interface—ten synchronized strips of 35 mm cinema film on which she drew. The link between vision and sound, which you might expect to be direct, is here obscured by the very idiosyncrasy of the instrument's design, which is now the subject of a joint research project between Goldsmiths College and the Science Museum.

This is just one instance of how electronic music provides clear examples of path dependency in technological/cultural change. This can also be seen in the development of novel musical forms: the BBC Radiophonic Workshop[7] and Terry Riley and Brian Eno working with Robert Fripp separately invented a new kind of music when they discovered they could create complex canonical soundscapes duetting with themselves when they ran the same piece of quarter-inch tape between two tape recorders, recording on the first and playing back on the second. The digital delay has replaced the reel-to-reel tape machine, and now this form of music is so familiar that a "loop pedal" is a standard guitar effects device, and "tape delay" is reproduced in digital audio workstations (DAWs) such as Ableton Live.

Path dependency also provides a valuable way of thinking about the "interfaces" of musical instruments, an important topic that is discussed in several chapters. Were new technical sound possibilities limited by conventional interfaces? How did the developers of new electronic instruments cope with this? Today we take it for granted that a synthesizer must have a keyboard, but Trevor Pinch and Frank Trocco have shown that this wasn't originally the case, and they explain how and why Robert Moog ended up with a keyboard for his synthesizers.[8] The broader picture is explored by Tellef Kvifte in this volume, as he looks at similarities and differences in the development of "user interfaces" of acoustic and electronic instruments, comparing the technology of the nineteenth century and the second half of the twentieth century. Using examples from the collection of Ringve Museum in Norway, he finds that in both periods, several new interfaces were developed. But while the first period had a strong focus on pitch control, the development in the second period was mainly concerned with timbral qualities.

If today all music is to some extent electronic, it is debatable whether the electronic revolution has been as rapid as it might have been if technology simply determined music. What we see in the history of music is a complex interplay among cultural forms, aesthetics, technical devices, and economics. The importance of electronic studios from the mid-twentieth century within the avant-garde should not be underestimated. Yet the comparatively late dominance of popular music by electronic devices may owe as much to the dramatic reductions in the cost of microelectronic devices over the last three decades as to any compelling aesthetic motivation. Perhaps, when Roland, Casio and Yamaha realized the potential of electronic keyboards to be viable consumer products, they kick-started the electronic revolution in popular music. Certainly, the availability of inexpensive digital synthesizers made it as cheap for amateurs to form an electropop band (two keyboards and a drum machine), as the conventional rock band (two guitars, a bass, and drums). This raises another enduring dynamic in the history of music, its instruments, and its technologies; namely, the relationship between amateur and professional musical activity. Here again, new technologies have consistently remapped this most permeable of distinctions. The player piano might have promised to replace the amateur pianist with the ghostly presence of a Liszt or an Elgar, but equally the 45 rpm single encouraged teenagers to buy an electric guitar and try their hand at skiffle or punk. Similarly, digital distribution of music via the Internet has provided opportunities as much for laptop-wielding bedroom electronic musicians as for sourcing that rare and forbidding masterpiece of the postserial avant-garde.

Examples such as these are explored in the chapters that follow. Some authors here show how particular localities produced particular technological and musical phenomena. Peter Donhauser introduces the reader to the musical collection of the Vienna Museum of Technology as he outlines the development of electronic musical instruments in Austria between 1920 and the late 1950s. Based on extensive documentation, he describes the construction, reception, and fate of instruments such as the Superpiano and the Magneton. However, the instruments are also examples of how the vast majority of early electronic instruments remained prototypes and never achieved commercial success. Finally, in the last chapter of this volume, David Toop treats another flourishing milieu during a different period; London's emergent improvised music and sound art scenes in the 1970s. In revisiting the circumstances of his 1974 text *New/Rediscovered Musical Instruments*, he traces connections between experiments in live electronic music, improvization, and the parallel growth among the musicians of that period of interest in ethnomusicology and the study of musical instruments. He shows links to museum culture that may well be unexpected to today's music enthusiasts.

In total, by showing the different contexts in which musicians, technicians, and consumers have adapted—and adapted to—changing musical instrumentation technologies, and by close examination of the instruments and devices themselves, these chapters help the listener to hear different musics afresh, the scholar to see music as a prime site for unraveling technology and culture, and the museum visitor to validate their own experience.

Notes

1. For example: Elena Ungeheuer (ed.), *Elektroakustische Musik. Handbuch der Musik im 20. Jahrhundert. Band 5* (Laaber, Germany: Laaber-Verlag, 2002); Hans-Joachim Braun (ed.), *Music and Technology in the Twentieth Century* (Baltimore, MD: The Johns Hopkins University Press, 2002); Peter Donhauser, *Elektrische Klangmaschinen: Die Pionierzeit in Deutschland und Österreich* (Vienna: Böhlau, 2007); IMA Institut für Medienarchäologie, *Zauberhafte Klangmaschinen. Von der Sprechmaschine bis zur Soundkarte* (Hainburg, Germany: Schott, 2008).

2. Including: Albert Glinsky, *Theremin: Ether Music and Espionage* (Champaign: University of Illinois Press, 2005); Jean Laurendeau, *Maurice Martenot: Luthier de L'Électronique* (Croissy-Beaubourg, France: Dervy-Livres, 1990); Trevor Pinch and Frank Trocco, *Analog Days: The Invention and Impact of the Moog Synthesizer* (Cambridge, MA: Harvard University Press, 2002); Gayle Young, *The Sackbut Blues: Hugh Le Caine, Pioneer in Electronic Music* (Ottawa: National Museum of Science and Technology, 1989).

3. Trevor Pinch and Karin Bijsterveld, "Sound Studies: New Technologies and Music," *Social Studies of Science* 34(5) (2004): 635–48. Trevor Pinch and Karin Bijsterveld, ed. *The Oxford Handbook of Sound Studies* (New York: Oxford University Press, 2011).

4. Steve Reich and Paul Hillier (eds.), *Writings on Music 1965–2000* (New York: Oxford University Press, 2004).

5. Ungeheuer, "Imitative Instrumente und innovative Maschinen?" in IMA Institut für Medienarchäologie, *Zauberhafte Klangmaschinen*, 45–59.

6. Louis Niebur, *Special Sound: The Creation and Legacy of the BBC Radiophonic Workshop* (Oxford, UK: Oxford University Press, 2010).

7. F. C. Brooker, *Radiophonics in the BBC (BBC Engineering Division Monograph 51)* (London: BBC, 1963), 7.

8. Pinch and Trocco, *Analog Days*, 58–62.

Bibliography

Braun, Hans-Joachim, ed. *Music and Technology in the Twentieth Century.* Baltimore, MD: The Johns Hopkins University Press, 2002.

Brooker, F. C. *Radiophonics in the BBC (BBC Engineering Division Monograph 51).* London: BBC, 1963.

Donhauser, Peter. *Elektrische Klangmaschinen: Die Pionierzeit in Deutschland und Österreich.* Vienna: Böhlau, 2007.

Glinsky, Albert. *Theremin: Ether Music and Espionage.* Champaign: University of Illinois Press, 2005.

IMA Institut für Medienarchäologie. *Zauberhafte Klangmaschinen. Von der Sprechmaschine bis zur Soundkarte.* Hainburg, Germany: Schott, 2008.

Laurendeau, Jean. *Maurice Martenot: Luthier de L'Électronique.* Croissy-Beaubourg, France: Dervy-Livres, 1990.

Niebur, Louis. *Special Sound: The Creation and Legacy of the BBC Radiophonic Workshop.* Oxford, UK: Oxford University Press, 2010.

Pinch, Trevor, and Karin Bijsterveld. "Sound Studies: New Technologies and Music." *Social Studies of Science* 34(5) (2004): 635–48.

———. eds. *The Oxford Handbook of Sound Studies.* New York: Oxford University Press, 2011.

Pinch, Trevor, and Frank Trocco. *Analog Days: The Invention and Impact of the Moog Synthesizer.* Cambridge, MA: Harvard University Press, 2002.

Reich, Steve, and Paul Hillier, eds. *Writings on Music 1965–2000.* New York: Oxford University Press, 2004.

Ungeheuer, Elena, ed. *Elektroakustische Musik. Handbuch der Musik im 20. Jahrhundert. Band 5.* Laaber, Germany: Laaber-Verlag, 2002.

———. "Imitative Instrumente und innovative Maschinen? In IMA Institut für Medienarchäologie," *Zauberhafte Klangmaschinen. Von der Sprechmaschine bis zur Soundkarte.* Hainburg, Germany: Schott, 2008.

Young, Gayle. *The Sackbut Blues: Hugh Le Caine, Pioneer in Electronic Music.* Ottawa: National Museum of Science and Technology, 1989.

Chapter 1

Bellowphones and Blowed Strings

The Auxeto-Instruments of Horace Short and Charles Algernon Parsons

Aleksander Kolkowski and Alison Rabinovici

Introduction

The quest for louder sound reproduction in the acoustic era (1877–ca. 1925) occupied the minds of the greatest inventors and engineers of the age.[1] It is no surprise, then, that the most successful and powerful mechanical device to amplify recorded sound—the Auxetophone—was the product of the combined efforts of Horace Short (1872–1917), of Short Brothers' aviation fame, and Sir Charles Algernon Parsons (1854–1931), inventor of the modern steam turbine engine. Compressed air was modulated by a valve, simulating mechanically the vibrating action of vocal cords, the blast of sound projected by a giant horn. When playing gramophone records, this reproducer could be used in the open air, in large halls, and stadia, and could even compete with an orchestra (Figure 1.1). Parsons's use of the valve on acoustic string instruments in the early 1900s resulted in the auxeto-cello and double-bass, with further experiments on the violin, piano, and harp—the very first externally amplified musical instruments, preceding electronic devices by decades.[2]

This chapter describes how, despite endorsements from leading musicians and scientists, and in stark contrast to the remarkable initial success of its gramophone counterpart, the new auxeto-amplified instruments met with open hostility from the musical fraternity and a mixed response from the press.[3] Parsons, increasingly disillusioned and

FIGURE 1.1
Charles Parsons's Auxetophone-Gramophone (ca. 1905). A hand-cranked Gramophone & Typewriter Ltd. machine is modified to carry the Auxetophone's tone arm with a counterweight, tubing, an air valve, and an oversized horn. The cabinet beneath conceals an air compressor. Courtesy of the Science Museum (Inv: 1938-313). Images available from Science & Society Picture Library.

in the red, abandoned his groundbreaking enterprise; it was nevertheless another decade before mechanical systems of amplification began to be rendered obsolete by the thermionic valve, microphones, and "wireless" technology.[4]

The auxeto-gramophone and its air-powered progeny, notably Harry Gaydon's "Stentorphone," have been only modestly documented in the history of sound reproduction technology and generally excluded from all but the most detailed descriptions and chronologies.[5] As few were ever commercially manufactured, these machines have also become much sought-after collectors' items. Nevertheless, they were at the very cusp of changing attitudes in twentieth-century mass culture and entertainment, when the shared experience of listening to recorded music was to become an acceptable and popular cultural activity.

Consigned to a footnote in biographies of Charles Parsons, neglected by historians and totally unexplored by organologists, auxeto-instruments remain important in any discussion about the mechanical amplification of sound and the early attempts to amplify live music. This chapter examines the Auxetophone method of instrument amplification from both organological and historical perspectives: a focus on the invention; its design, function, and construction, is complemented by a discussion of the social and cultural context in which this mechanical stentor was created. Parsons's air-powered auxeto-violin (patented in 1903) is also examined in the light of earlier and contemporary endeavors to increase the loudness of sound reproduction devices and instruments.[6]

Of particular importance is the diaphragm and horn-amplified violin of John Matthias Augustus Stroh (patented in 1899).[7] Parallels are made between the ways in which both Stroh and Parsons introduced their new devices to the public. Both used the scientific forum—the "conversazione" or the public lecture theater—arranged for public demonstrations in more popular venues, and both used the scientific and popular press to promote their inventions. By integrating research across both scientific and musicological disciplines, new conclusions have been drawn about the significant part played by the conductor Henry Wood in the use of auxeto-instruments during the 1906 Queen's Hall Promenade concert season.

The Auxetophone and its application to musical instruments represented a radical shift in the function and reception of recorded sound and amplified music, making it possible for the first time in history to successfully play at significantly louder volume in public spaces. The Auxetophone is the precursor to the public address system and electrical pickups for musical instruments that have so transformed modern music making.

The name *Auxetophone*, combining the Greek terms for "increasing" and "voice" or "sound" was suggested by Professor George Johnstone Stoney, esteemed mathematician, physicist, and mentor to Charles Parsons. Stoney made an attempt to analyze the quality of tone the device produced, calculating the remarkable and effective reinforcement of the upper partials that was a defining characteristic of the Auxetophone "sound."[8]

Nineteenth-Century Noise

Long before the invention of sound reproduction and sound amplification technology, unregulated music in public spaces was regarded as noise, and therefore as a public nuisance. Indeed, legislation was introduced in Britain in 1864 specifically intended to address the problem.[9] Throughout the 1850s and 1860s, the "tyrannical power" of street musicians and organ grinders and their intrusion into the private realm was discussed almost weekly in the British press.[10] Little wonder, then, that the advent of sound recording, or more precisely, sound reproduction, caused in some minds, at least, anxieties about what was early on seen to be a fundamental change in the ways in which society functioned and the ways in which social space would be utilized.

By 1878, anxieties had crystallized around Thomas Edison's "aerophone."[11] In somewhat hyperbolic language, the *New York Times* described it as a modification of a phonograph that "converts whispers to roars"[12] (Figure 1.2), while *Figaro* hypothesized about "a steam machine which carries the voice a distance of one and a half miles . . . You speak to a jet of vapor. A friend previously advised can answer you by the same method."[13]

It is uncertain if Edison ever made a fully functioning aerophone; one reporter's inquiry solicited the response:

> I haven't time to attend to that, I'm so busy with the phonograph. The aerophone is very simple. It isn't like a calliope, which requires a keyboard and different notes. It has only one note, and the vibrations of that are formed into words by the escape of the steam.[14]

Edison's 1878 British patent for the tin foil phonograph describes the aerophone principle more vividly:

> The sound vibrations from a phonet [a mouthpiece or a phonograph sound reproducer] are conducted by a tube . . . to (a) diaphragm that controls a valve in a tube connected with a reservoir of air or other fluid under pressure, and the air as it escapes by the valve passes to the trumpet-shaped end of the tube and produces sounds that are very loud and clear.[15]

The wildly exaggerated potential of this instrument to make speech audible within a radius of four miles would result, continued the *New York Times*, in the "complete disorganization of society."[16] That same year, imaginations had also been fired by David Edward Hughes's new carbon microphone.[17] Demonstrated by Hughes to be sensitive enough to register the footfall of a fly, the imagined potential for this contraption to amplify miniscule sounds to the level of a trumpeting elephant or a herd of horses gained general currency in the contemporary press; one article even appeared under the title "The Language of Flies."

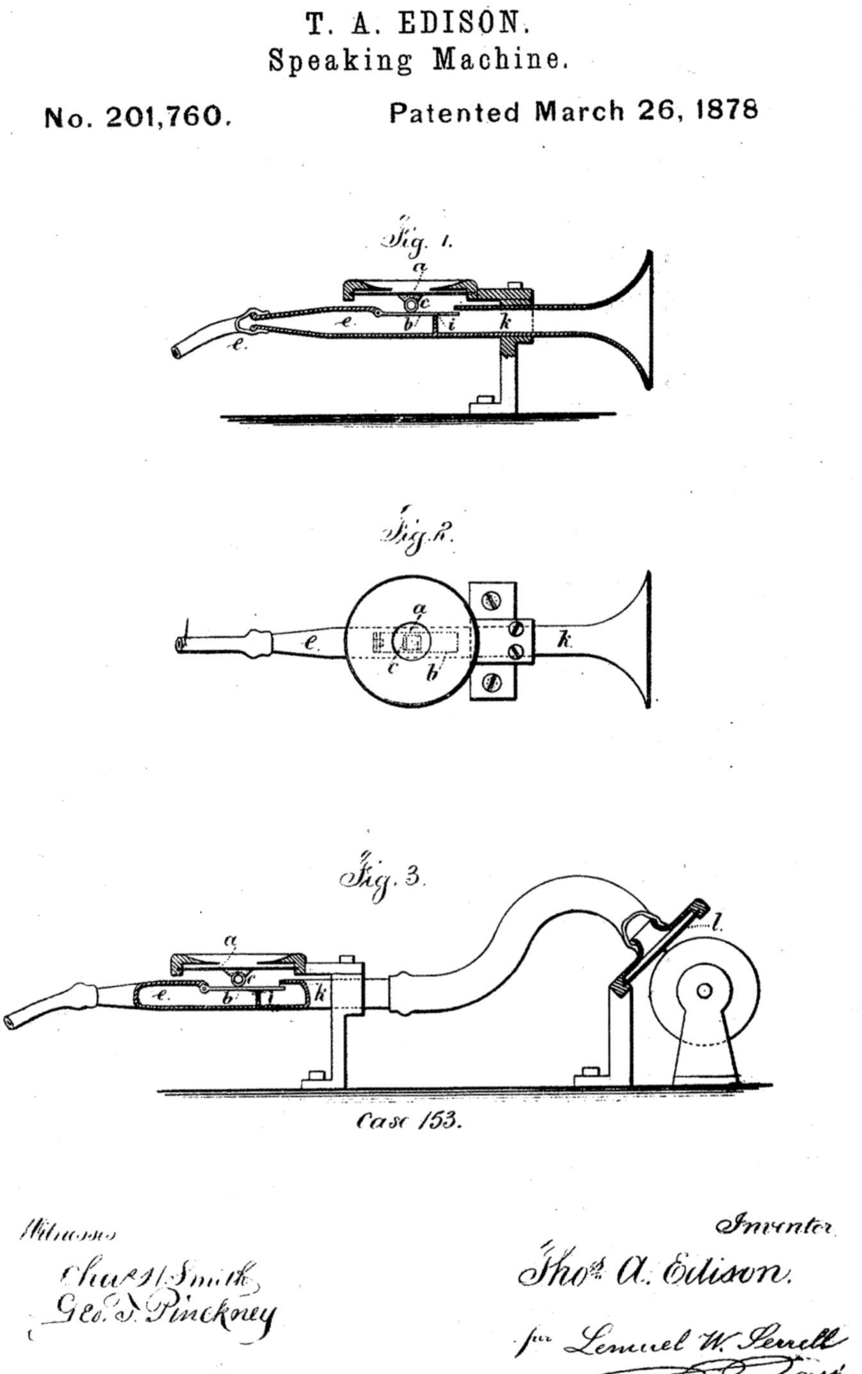

FIGURE 1.2

Thomas Edison's 1878 U.S. patent for a speaking machine commonly known as the *aerophone*. Figures 1 and 2 show the diaphragm *a* that is activated by the human voice or other sounds, connected to a pressurized air valve *b*; figure 3 shows the device as used to record on the surface of a cylinder, the amplified sounds increasing the vibrations of the recording diaphragm *l* and subsequent indentations made by the attached cutting stylus.

> There appears to be a scientific craze just now for the repetition of sound, or, in other words, increasing the noises of the earth. After the telephone and phonograph come the microphone, which magnifies sound and aids the ear as the microscope does the eye.[18]

The microphone took half a century to evolve beyond being principally a telephone apparatus. It had to wait until the mid-1920s before it could successfully be used for broadcasting and recording.[19] Thomas Edison's principle of sound magnification through compressed air, though, would be realized somewhat sooner.

The Magnificent "Gouraudphone"

By the end of the nineteenth century, mechanical sound reproduction was commonplace, and if not always welcome, it no longer precipitated the same degree of anxiety. Indeed, much energy and invention was invested in finding ways to increase the amplitude of recorded sound—methods including friction wheels, relays, and double diaphragms.[20] But there was no viable alternative to the flat, or later, coned stiff diaphragm system until 1898 and the arrival of pneumatically powered sound.[21]

While biographical and technical literature surrounding the invention and commercial exploitation of what was to be known as the Auxetophone are closely associated with Charles Parsons, it was another engineer, Horace Short, who in 1898 developed and patented the first working models that bear a striking similarity to Edison's aerophone system. The association with the Wizard of Menlo Park continues: Short's new sound magnifier, the "Gouraudphone," was named after, and developed with, the financial backing of Colonel George E. Gouraud. This charismatic entrepreneur was Thomas Edison's close associate, his European agent and publicist for the phonograph, and responsible for some of the most significant early recordings of speech and music.[22] A technophile and fanatical promoter of state-of-the-art communications, Gouraud had some thirty years earlier aided Edison in finding investment to develop telegraph and telephone technology. In doing so, he inadvertently aided the invention of the phonograph. Gouraud was now to promote and invest in Horace Short's experimental amplification device with the same vigor he had shown when supporting Thomas Edison.

Short had seen a phonograph while traveling in North America during the 1890s and began to formulate his own ideas around sound reproduction using compressed air.[23] It is no coincidence that he should subsequently develop an air-powered amplification system in Mexico while working as a mining engineer, as compressed air was commonly used in mining for drilling machines and other applications.[24]

Short returned to England in 1898, and after meeting with Colonel Gouraud to secure finance for his "mechanically enhanced megaphone system," applied for a patent that same year.[25] Although it can be seen simply as a variation to Edison's 1878 aerophone, Short's patent was nevertheless issued in 1899.[26]

Short's innovation was to introduce a hinged, spring-loaded grid or comb-type valve consisting of minute "tongues" that opened and closed a series of slots on a fixed grid to control a flow of air under pressure. As with the aerophone, this opening and closing of the air valve was governed by a flat diaphragm mounted on a mouthpiece and set into motion by a human voice. As both diaphragm and hinged grid are connected by a spindle, the vibrations of the diaphragm and the opening and closing of the valve operate in unison, causing corresponding oscillations in the column of air that is forced through the apertures of the grating at extremely rapid intervals and discharged from a horn. Alternatively, a phonograph reproducer could be used to control the air valve instead of a voice diaphragm so as to play recordings at great volume. Short uses the term *telephone diaphragm* (Figure 1.3) in his further patent of 1901,[27] and a comparison may be drawn in the way that the diaphragm modulates a current of air in a Gouraudphone as it does in a telephone to a current of electricity.[28]

The first public trials took place in the picturesque landscape of the South Downs. Short's workshops were close to Devil's Dyke, a popular scenic destination, and in July 1900, locals and sightseers were subjected to a barrage of loud recorded announcements from the roof of Short's "Menlo laboratory,"[29] where he had set up his Gouraudphone augmented by a four-foot horn.[30] Reactions were mixed. Local newspapers named it the

FIGURE 1.3
Horace Short in 1900 posing with his newly invented Gouraudphone. Courtesy of Brassey's: Putnam Aeronautical.

"Howling Terror," fearful that its "diabolical broadcasts would dominate the forthcoming century," while the *London Mail* enthused:

> The possibilities of the machine are practically endless. It will render loud selections in the open air that can be listened to by thousands of people, or it will shout news messages that could be heard high above the roar of the traffic and the thousand noises of a big city.[31]

Colonel Gouraud claimed that the instrument was capable of making the human voice audible over enormous distances, some "tens of kilometers." His vision for its use as an "oratorograph"—for political, musical, or literary purposes—foresaw public address systems, while it could also function effectively as an articulating siren for ship-to-ship communications.[32]

The Cirque à Boulogne was the chosen venue for the second public demonstration in August 1900, but part of the apparatus broke in transit, allowing only for selections using an adapted phonograph, and not live sound, as intended. Cylinder recordings of speech, songs, military trumpet calls, and cornet solos were reproduced through the "gigantic graphophone funnel . . . with twenty-fold strength."[33] Gouraud then took Short and the device to the Paris Exposition Universelle in September 1900, where they set up the Gouraudphone on the top of the Eiffel Tower. Their purpose was to project music and speech at a level of amplification that would reach beyond the fortifications of Paris.[34] The Exposition was the largest exhibition of its kind ever seen, with as many as fifty million visitors and over 76,000 exhibitors.[35] Perhaps it was because of the huge melee of competing exhibits that the press seemed largely to have ignored the Gouraudphone demonstration. However, Horace Short's brother, Oswald, wrote in 1918 that "records were (afterwards) roared out by the instrument from the top of the Eiffel Tower where we had the use of Eiffel's room . . . being about 1,000 feet from the ground" and that the "Air Compressor which was used with the instrument" was successfully driven by liquid air.[36]

Despite Colonel Gouraud's best efforts, his energy in promoting the device, and publicity gained from the public demonstrations, he was unable to secure investment. Facing financial difficulties and after losing a protracted litigation with the landlord of the Menlo Laboratory, Gouraud shut it down.[37]

Horace Short, who had relied on Gouraud for finance, was now out of work, in debt to his patent agents, and in dire need of a sponsor. His luck was to change dramatically through the intervention of one of the foremost engineers of the age, Charles Parsons, who had himself developed a sound-magnifying air valve in 1900. Parsons actively sought to gain exclusivity to the system, and in 1904, a deal was struck with Short, who sold his 1898 patent rights to him for £700. Short entered into Parsons's employment in order to work exclusively "on the system to which our patents refer." He added, "I don't think that you will regret any step you may take to meet this end."[38]

Parsons had already registered two patents for a compressed air amplification device, the first in 1903 and the second in 1904.[39] He demonstrated the device at the Royal Society in May of that year. He later claimed to have been unaware of Short's work and had acted only after Short attempted to enforce his patent on hearing the publicity surrounding the Royal Society demonstration.[40] The truth was that Parsons had discovered Short's prior grant when filing his own applications, and he had been trying to obtain Short's British and overseas patents through the services of an intermediary in 1903.[41]

Short remained in Parsons's employment until 1908, when he left to make his fortune designing and building aircraft with his brothers Oswald and Eustace, establishing the world's first aviation company. He died in 1917, at the age of forty-four. What remains clear is that Short's role in the development of the Auxetophone was far more significant than the current Parsons hagiography has allowed.

Industrial Strength Sound: Charles Algernon Parsons and the Auxetophone

Convinced he could improve the sound quality of music reproduced by gramophones and phonographs, Charles Parsons had, in 1901, begun to experiment using controlled air under pressure.[42] He used the organ pipe and mechanically operated siren as models rather than the vibrating diaphragm and soundbox principle, with its inherent limitations of volume and relatively poor characteristics of tone. What began as a hobby soon became a serious business enterprise. Parsons's commitment to the air valve was unwavering, even at a time when he was also deeply preoccupied with huge-scale business projects, including the design of turbo alternators for the Wallsend-on-Tyne power station, and steam turbines and turbine engines for the *Mauritania*, *Lusitania*, and H.M.S. *Deadnought*.

Perhaps it was Lady Parsons herself who most eloquently documented her husband's passion for the project. After Parsons died in 1931, one obituary quoted from Lady Parsons's memoirs:

> We always had a workshop in our house where Charles spent most of his time at home, working till 2 or 3 a.m. The most trying time for the family was when he was producing the "Bellowphone." Strange and weird were the noises through the nights. The finished Bellowphone was a very sweet and beautiful instrument when played by him at home, with the sound coming through a gigantic trumpet. He used to place it in the garden, and people from miles round came flocking into our park to hear it.[43]

Of Parsons's two patents jointly registered in 1903, the first, for Improvements in Sound Reproducers or Intensifiers Applicable to Phonographs, Gramophones, Telephones and the Like shows a valve designed to control a flow of compressed air with a comblike movable grating system very similar to Short's. Both versions employ a fixed grid and a movable one, which opens and closes the slots in the former through which the air is forced out. The only significant difference from a Gouraudphone (or aerophone) is that the valve is controlled directly by the movements of a gramophone needle as it rides the groove of a disc record or a phonograph stylus on a cylinder and not by a voice-activated diaphragm or phonograph reproducer. Pulsations of air from the valve's rapid opening and closing correspond directly to the undulations in the record groove, creating sound waves identical to those originally recorded. A further improvement filtered the air through a gauze frame and cotton wool to ensure that no particles would get stuck in the grating and interfere with the workings of the valve (Figures 1.4 and 1.5).

The second patent was for Improvements in and Relating to Musical Instruments (Figure 1.6, bottom). Here, the same valve is adapted for use with string instruments such as the violin, cello, double-bass, and also the harp and piano. Using a violin for his model, Parsons replaces the soundboard and resonating body of the instrument with an air valve operated by vibrations of the strings. The connecting rod to the movable

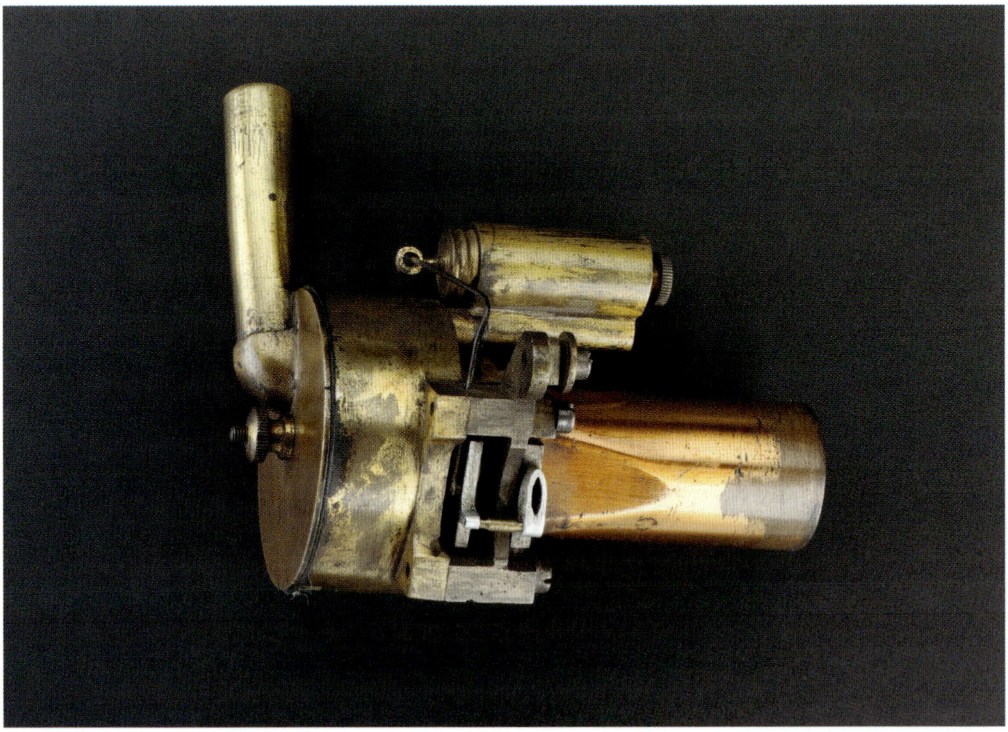

FIGURE 1.4
Prototype Auxetophone air valve for gramophone (ca. 1904). It has a "compensating cylinder" to adjust sensitivity and to guard against fluctuations in air flow. Courtesy of the Science Museum (Inv: 1938-313). Image available from Science & Society Picture Library.

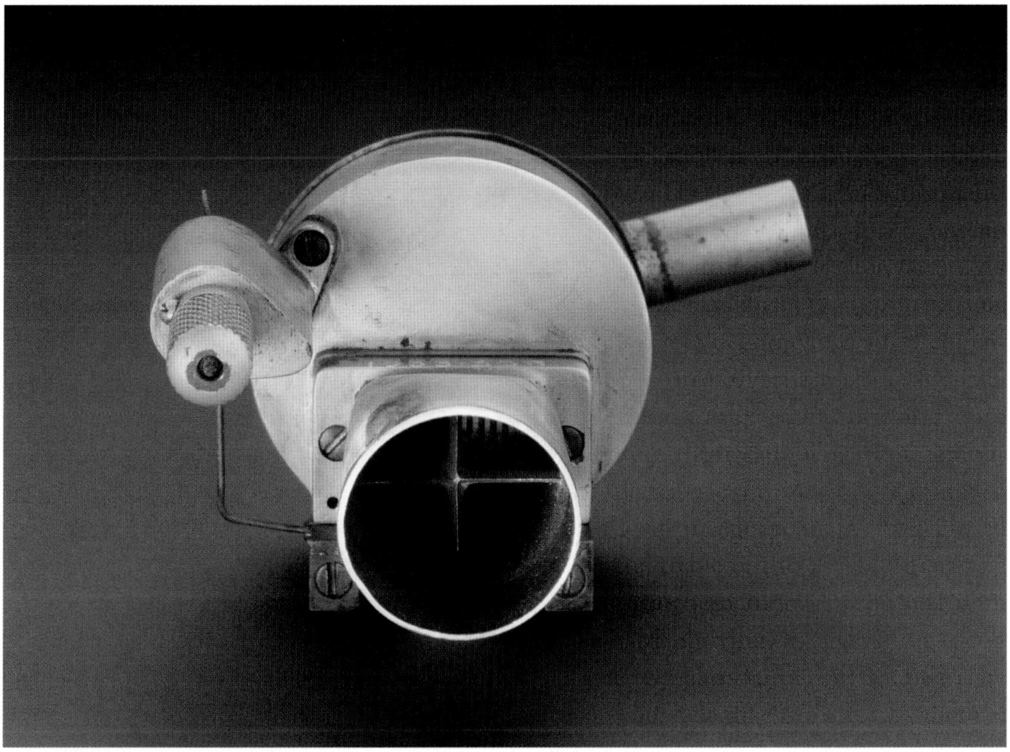

FIGURE 1.5
Improved Auxetophone air valve for gramophone (ca. 1905). Courtesy of the Science Museum (Inv: 1938-358). Image available from Science & Society Picture Library.

grating, in this case, is directly attached to the violin's bridge, and as the strings are bowed or plucked, the side-to-side movements of the bridge open and close the valve. The rod consists of two tightly fitting concentric tubes that are lubricated with thick oil or fat in order to allow a slower relative motion between the bridge and the movable grating, but absolute rigidity for oscillatory motion. This had the effect of reducing some of the higher frequencies, and the use of this form of proto-low-pass filtering would be later covered in a patent as applied to gramophones.[44] The sound is discharged through an expanding horn attached to the air valve's exit side. In what could well be the first patented noise reduction technique, Parsons lines the horn with velvet, "which has the effect of damping out scratching sounds and very high harmonics."[45]

Contemporaneously, air was used to activate the keys of player pianos and had previously been made to sustain the notes of experimental keyboard instruments such as the anémocorde and piano éolien.[46] Until now though, a compressed air system had not been devised that could amplify the vibrations of a string instrument to such magnitude. Parsons claimed that "the phase of motion of the strings are much more truly and accurately reproduced in air waves" and that "the power and character of the sounds is greatly superior to any ordinary instrument."[47]

Parsons's second patent for improvements to musical instruments must be regarded as directly inspired by the work of Augustus Stroh (1828–1914). Stroh had registered a patent for a mechanically amplified violin (known as the Stroh violin) in 1899 (Figure 1.6, top), and by 1904, the Stroh instrument was in regular use in the string sections of recording studio orchestras and by soloists.[48] If one ignores the supply of compressed air, there are marked similarities between the two new mechanically amplified violins. Both are solid-body instruments with expanding horns; both have an adjustable rod connecting the bridge to the valve or diaphragm. In the case of the Stroh violin, the bridge sits on a rocking lever turning on a knife edge; this lever is joined to the rod, which pushes and pulls the large aluminum diaphragm in accordance with the bridge movements.[49]

By applying the air valve to all forms of musical amplification, both live and recorded, Parsons was throwing down the gauntlet to the music industry, challenging conventional practices of sound reproduction in the same bold manner he had shown a few years earlier humiliating the Royal Navy with his steam-turbine-driven vessel *Turbinia*.[50]

Parsons's compressed-air amplified violin was introduced to the members of the Royal Society at a Conversazione in May 1904. Not however, as a newly constructed instrument, it is more likely that an air valve was attached to a conventional fiddle. "The exhibits were . . . numerous and varied" wrote *The Times* reviewer. "The Hon. C. A. Parsons' Auxetophone is a very ingenious arrangement consisting of an air operated valve, to be used for a reproducer in gramophones or phonographs, in the place of the usual reproducing diaphragm. Its application to the violin, as exhibited in the lecture-room, afforded a very interesting and striking demonstration."[51]

Not all reactions were favorable, however. An Edinburgh journal saw fit to ask: "Have you heard the auxetophone? It is to be hoped not. All Mr. Parsons' turbines will be wanted to take long-suffering humanity out of earshot of his diabolical invention."[52]

A further demonstration took place the following year at the rooms of Metzler and Co., one of London's leading musical instrument firms. The valve had been redesigned to incorporate new features and was applied to a gramophone. Great claims were made for the volume the instrument could produce; sound could apparently carry two or three miles, and the quality of the sound was improved.[53] By now, Horace Short was working full time on perfecting the Auxetophone, and he assisted in the "Metzler's Rooms" demonstrations.[54] Parsons had previously registered two further patents, which contained important modifications to the air valve. The first dealt with the connection of the needle and movable valve grating, as described in his earlier patent for musical instruments. The method impeded high-frequency vibrations that corresponded with needle-scratch and record-surface noises by forcing the linking rod to pass through a mixture of bicycle cement and castor oil. This resistive but elastic substance was said to give a more uniform tone.[55] In the last of Parsons's patents for the Auxetophone, Improvements in and Relating to Sound Reproducers, he added a small "compensating cylinder" fed by the main

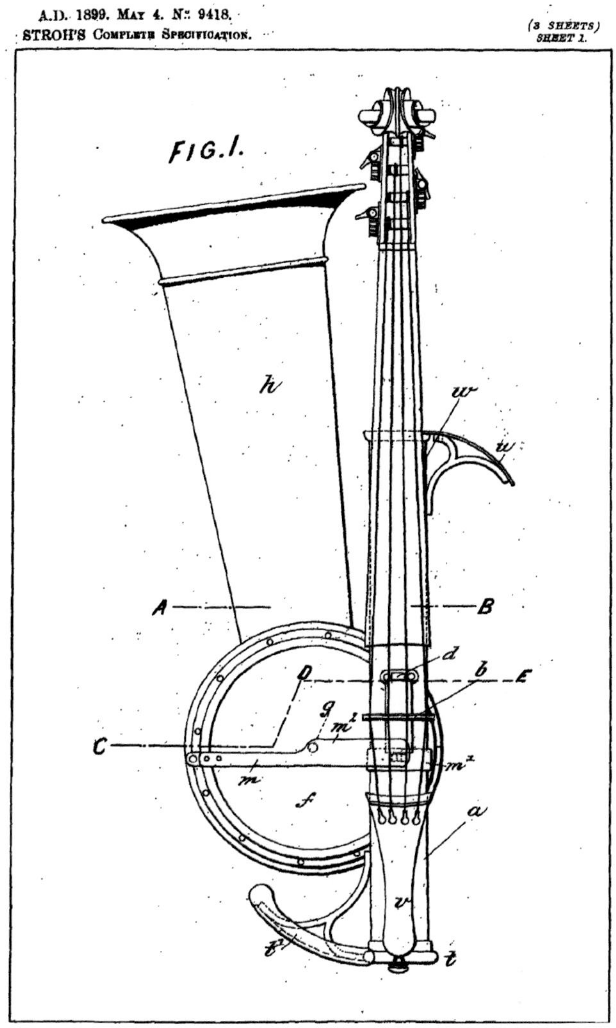

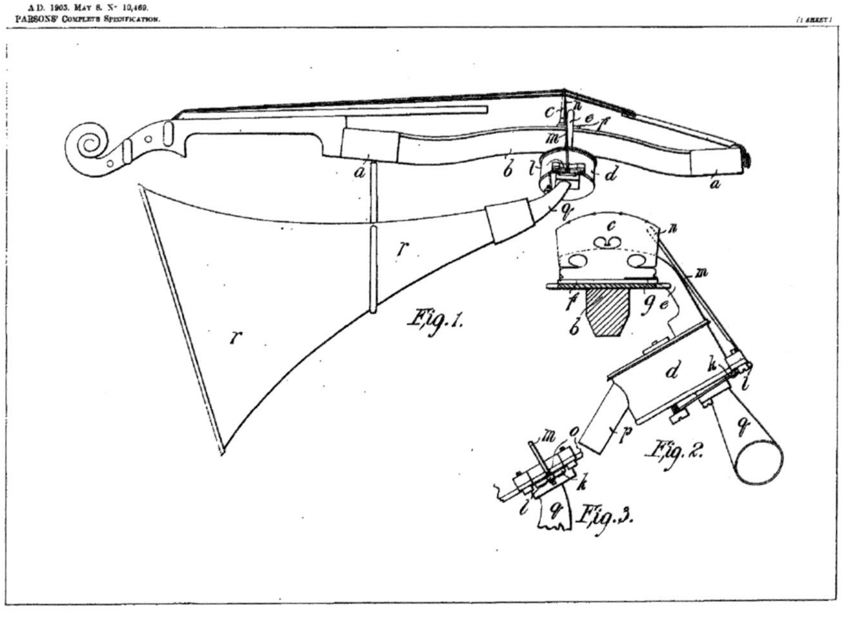

FIGURE 1.6
Two revolutionary string instrument designs: (top) Augustus Stroh's 1899 patent for a mechanically amplified violin and (bottom) Charles Parsons's 1903 air-assisted model.

air supply and containing a spring-loaded piston connected to the valve by a wire.[56] This enabled fine adjustment of the distance between the moveable comb and the fixed grid or valve seat, and it also helped to prevent fluctuations in the air supply from affecting the workings of the valve.[57]

The air pressure supplied by an electric pump could vary between two and four pounds per square inch, though two-and-a-half was found to be adequate (Figure 1.7). On one occasion at a London Earls Court trade fair in 1906, up to seven psi was erroneously used to demonstrate the Auxetophone with the result that the unfortunate spectators in the front rows left for the exits with their hands over their ears.[58]

Parsons's choice of a disc-playing Berliner gramophone for his demonstrations rather than a cylinder phonograph is as significant as it is puzzling; if his aim was to improve the quality of reproduced sound, then the phonograph cylinder would surely have been a more suitable medium. In 1904 the gramophone was still in its infancy, and compared to the phonograph, it suffered from inferior sound quality and distortion resulting from constant angular velocity.[59] Moreover, as Short had demonstrated using a recordable silver-coated cylinder in 1900, Parsons's device could also benefit from the increased amplitude a hard-surfaced cylinder would have provided, as well as allowing for self-made recordings.[60] However, using the gramophone did turn out to have been a shrewd maneuver, as the flat-disc record soon became the predominant sound-storage medium (Figure 1.8).

Just before the Metzler & Co. demonstration in March 1905, Parsons had sold the patent rights of the Auxetophone to the Gramophone and Typewriter Company for £5,000.[61] However, he retained the rights in relation to musical instruments, and Horace Short was set to work at perfecting the air valve for this purpose.

The Victorious Music Machine

A commercial version of the Auxeto-Gramophone was made available to the public in November 1906, after months of development by the Gramophone Company, Ltd., at Hayes, Middlesex, in tandem with the Victor Talking Machine Company in New Jersey (Figure 1.9), who also produced a model for the U.S. market.[62] The new Auxetophone sold for £100 in England, or $500 in the United States, making it affordable only to institutions, businesses, or wealthy audiophiles.[63] Designed "for large residences, ballrooms, hotels, theatres, halls, piazzas and lawns," it was declared that "no space is too large for a perfect rendering of Grand opera, Concert or a Dance Programme."[64]

Public space was never to be the same again, as the Auxetophone provided for the first time a successful means of projecting recorded sounds to mass audiences. The *Illustrated London News* forecast in May 1905 that:

> before the summer is over, the people who throng to our parks and summer pleasure resorts of evenings may be privileged to hear the voices of all the great singers of the world and the finest orchestral productions, as well as the voices and words of

FIGURE 1.7
Interior of Parsons's Auxetophone-Gramophone cabinet revealing the electrically powered air compressor. Courtesy of the Science Museum (Inv: 1938-313). Images available from Science & Society Picture Library.

FIGURE 1.8

The Auxetophone's air valve primed with a steel needle to play a disc record. Courtesy of the Science Museum (Inv: 1938-313). Images available from Science & Society Picture Library.

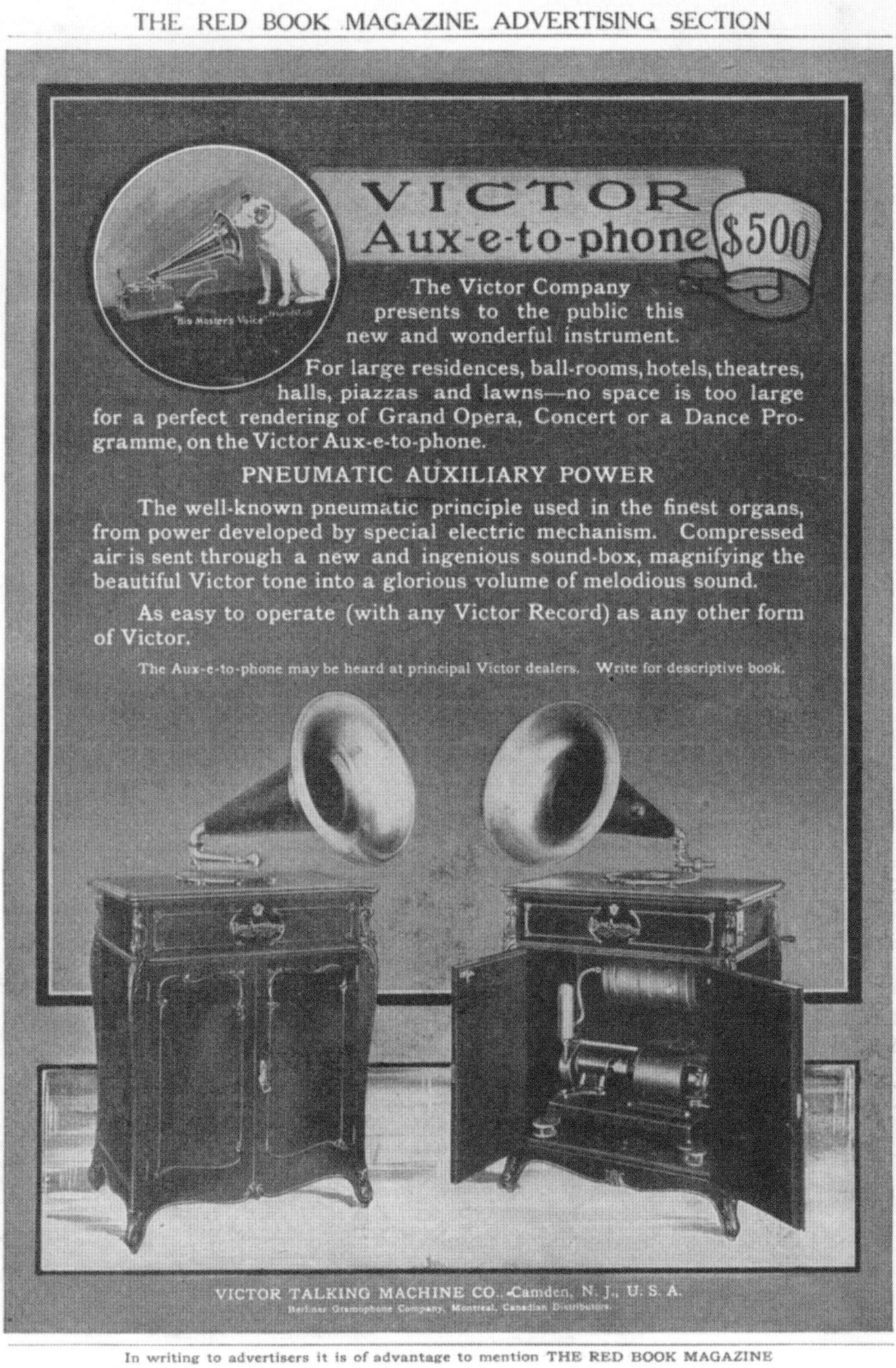

FIGURE 1.9
Advertisement for the Victor Auxetophone, 1907 (Author's private collection).

> many great and distinguished persons they may never chance to see. All this is to be brought about—not by the much maligned phonograph—but by the gramophone fitted up with a special attachment called an "auxetophone" which increases the volume of the gramophone and enables it to sound as loud as a full brass band in the open air.[65]

Not only was the result of the election for the governorship of New York in 1906 broadcast to the populace by means of a searchlight flashing from the top of *The Times* building, but the Auxetophone was used to provide additional entertainment. From a height of four hundred feet above the pavement:

> the novel feature of the evening will be a concert given by the Auxetophone . . . the last word in improvement of the phonograph. It has been specially devised to supply music to just such great gatherings as will assemble in Times Square on Election Night. It magnifies the sound in marvellous fashion and throws the notes to a great distance. National and popular airs will be played, and the gathering will be invited to sing to the accompaniment of the instrument.[66]

Thus was born a new form of popular engagement with technology. The popularity of gramophone recitals that were organized in parks, concert halls, and stadia is attested by contemporary press reports, photographs, and postcards showing huge gatherings listening to music broadcast from a single giant gramophone horn (Figure 1.10).[67] An

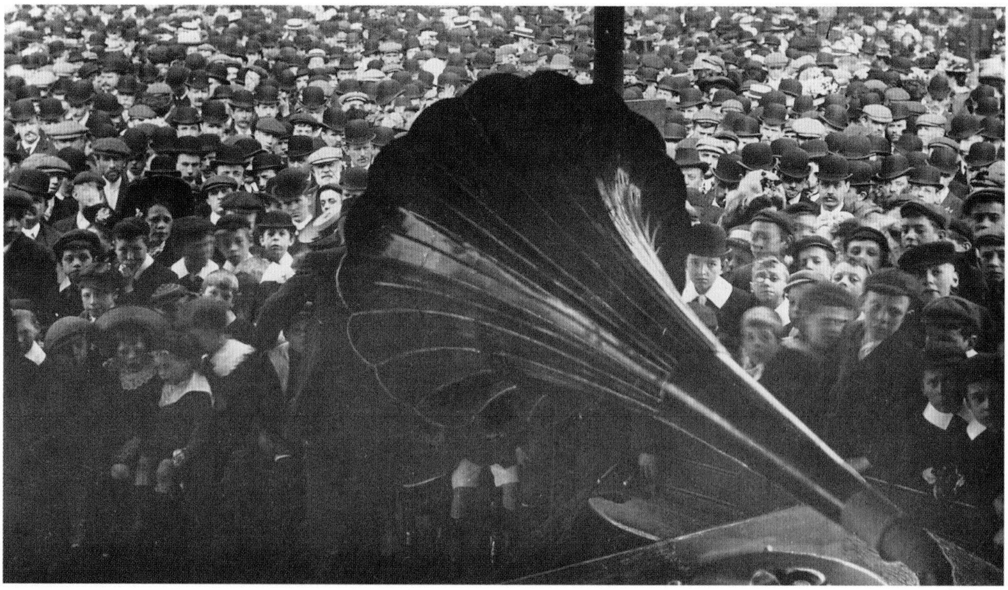

FIGURE 1.10
Postcard (ca. 1906): Queen's Park Manchester, "The Crowd Listening Attentively to the Auxeto Gramophone." Image courtesy of Patrick de Caluwé.

advertisement by the Gramophone and Typewriter Company claimed that an audience of ten thousand was entertained by the Auxetophone at the Albert Hall in 1906, with as many as forty thousand at the Crystal Palace on Empire Day 1911.[68]

In a remarkable echo of modern conventions in technologically based music performances, Auxeto-Gramophone recitals were often fully fledged audiovisual events, incorporating large-scale projected images of the artists being reproduced on record. A contemporary newspaper article from 1908 reported that:

> the [auxeto-] gramophone is all but perfect in the reproduction of the human voice—perhaps, indeed, perfect—but wanting only the fascinating presence of the sweet singer in the flesh. The fine full-length portraits thrown on the screen by limelight effect helped to heighten the delusion considerably.[69]

Playing with Recordings and Comparison Concerts

Reportedly as early as 1906, the Auxetophone was combined with instruments and voices.[70] Recordings of singers or solo instruments were accompanied by live instruments, and the merits of the Auxetophone's reproduction exposed to scrutiny. "Record and reality harmonize well and those dangerous pauses are well filled," wrote *The Times* reviewer in 1909,[71] while the live-organ accompaniment of a recording of the baritone Emilio de Gogorza in the Royal Albert Hall, was considered "miraculous."[72] So popular was this novel juxtaposition that the Victor Company published a series of twenty-four scores titled *Orchestrations* to accompany records of operatic arias and popular songs on its prestigious Red Seal label. Each score had a designated record of a famous opera singer such as Enrico Caruso, Nellie Melba, and Antonio Scotti, usually already backed by an orchestra. The arrangements were made for an ensemble of strings, flute, clarinet, cornet, trombone, and piano accompaniment (Figures 1.11 and 1.12).[73] The house bands of large hotels found themselves playing along to recordings as owners vied for trade, "adding the Auxetophone to their orchestras" in order to stage concerts featuring the most famous artists of the day—albeit in absentia.[74]

Writers and historians often cite the Edison Company's Tone Tests, many hundreds of which took place between 1915 and 1926, as the first serious attempts to directly compare recordings with the sound of live music, using state-of-the-art reproduction technology that was up to the task.[75] However, direct comparisons had taken place a decade earlier, from the very first public demonstrations of the Auxetophone in 1906 at the Albert Hall and elsewhere. The musical renditions of live soloists were repeated by the Auxetophone in succession, and audiences were amazed to hear, in their judgment, "no perceptible difference."[76] Since the invention of the phonograph, recorded music had commonly been treated with scorn and suspicion by music critics and the concertgoing public, with particular vitriol aimed at the quality of reproduction. With the arrival of the Auxetophone, the activity of listening to concerts that featured recorded music was to gain cultural respectability.[77]

Orchestration for

Victor Record No. 88127

Celeste Aïda (Heavenly Aïda)

Aïda

(Verdi)

Sung by Caruso

Price $1.00

Arranged for

Conductor and 1st Violin,	Cello,	Clarinet,
2d Violin,	Bass,	Cornet,
Viola,	Flute,	Trombone,

Piano Accompaniment

INSTRUCTIONS FOR USE OF THE VICTOR AUXETOPHONE IN CONNECTION WITH ORCHESTRA

I. See that the instrument is wound up completely before playing each record.

II. Allow the turntable to revolve five or six times, so that it may attain its full speed, before attempting to get the proper pitch or to commence playing.

III. How to get the pitch—

- *a.* Set the Speed Indicator at 76 (which indicates the number of revolutions of the turntable per minute.)
- *b.* Play the first measure of the orchestration with some one instrument, and
- *c.* At the same time play the first measure of the record on the Auxetophone. This will allow you to determine whether or not the Auxetophone is playing at the proper pitch.
- *d.* If the pitch is too low, increase the speed by moving the indicator to 78 or higher, until you find the correct pitch; if too high, move the indicator to 75 or lower, if necessary. Then test with orchestra instrument as before.
- *e.* Should the correct pitch of any particular selection be found at a speed other than 76, mark the correct speed on the record for future reference, thereby doing away with the necessity of finding the pitch each time a selection is re-played.

IV. To determine whether or not the turntable is revolving at the correct number of revolutions per minute, as indicated on the dial, place a piece of white paper under the edge of the record, and count the revolutions.

Use a New VICTOR needle for each record played.

FIGURE 1.11

A Victor "Orchestration" for live accompaniment to an Auxetophone-reproduced record. Image courtesy of Patrick de Caluwé.

FIGURE 1.12
Conductor's score from a Victor Auxetophone "Orchestration." Image courtesy of Patrick de Caluwé.

Henry Wood and the Amplified Bass

Charles Parsons hoped to mirror the success of his air valve as used on the gramophone by applying it to musical instruments. With Horace Short engaged at Heaton Works, they soon began to experiment with the device on an entire string section. In November 1905, before the Northern Scientific Club in Newcastle-on-Tyne, Parsons gave a demonstration, and again claims were made not only for the volume but also for the quality of sound produced. The *Musical Times* wrote:

> Selections were played on two violins, violoncello, and double-bass, singly and in concert. When the string quartet was being played, the volume of sound was equal to that produced by a large body of strings, even though the trumpet-mouths were turned away from the audience.[78]

The conductor, Henry Wood, may well have witnessed the Newcastle-on-Tyne demonstration, and it was he who most famously utilized auxeto-amplified instruments in his orchestra during the 1906 season of Queen's Hall Promenade Concerts (Figure 1.13). Wood wholeheartedly endorsed the system, writing to Parsons that he had "tested it in all classes of music" and was "delighted with its real practical value as a new voice in the orchestra."[79]

Mention of Henry Wood's encounter with the Auxetophone is almost entirely missing from biographical accounts of his life.[80] *The Strad*, however, did follow developments with some interest. In August 1906, a contributor noted that the Hon. Charles Parsons "shewed to Mr. H. J. Wood and a few scientists, an invention of his by which the sound of violins, violas, violoncellos or double-basses could be so increased that one instrument would equal a dozen in volume of tone." The author objected to this possibility on the grounds that the individuality and character of the player would be lost in performance. He also expressed his objections to the consequent reduction in player numbers:

> What makes an orchestra so impressive in tone-quality is the playing together of a number of distinct and different personalities. If the string band is to be reduced in numbers, and the few performers left provided with artificial tone-producers the effect may be the same in sound impact, but otherwise I believe vastly different and inferior.[81]

Henry Wood, undeterred by such skepticism, had an Auxetophone valve fitted to a double bass in the Queen's Hall Orchestra within a month. Although Wood had witnessed Parsons's earlier Royal Society demonstration of an auxeto-violin, he did not use it in the 1906 Promenade season. As A. Q. Carnegie later noted in 1934, the application of the Auxetophone to the violin had not been satisfactory as the valve was too cumbersome. Carnegie also noted that the auxeto-amplified basses helped to achieve a balance

FIGURE 1.13
Auxeto double-bass. Image courtesy of Constable & Robinson.

of heavy stringed instruments that would have been impossible otherwise, owing to lack of space.[82] *The Strad* reviewer, after hearing the amplified bass in the orchestra, had not changed his mind about the "real musical value of the invention." Improvement in volume of the bass section, he believed, lay in the engagement of extra players, rather than in mechanical amplification of the instruments, clearly echoing the sentiments of orchestral musicians who feared for their employment.[83]

The arrangement in the Queen's Hall must have been truly impressive. Two giant horns were installed in the auditorium, and the compressed air was electrically pumped up from the basement.[84] The Queen's Hall management found it necessary to assure its patrons that the "trumpets resembling ventilating shafts on an ocean steamer were not part of an improved system of ventilation." Furthermore, management also complained about the cost of electricity needed to operate the air supply.

Henry Wood was highly enthusiastic, and he wrote to Parsons that he would "miss it terribly after the Promenades."[85] But the facts reveal pressing economic, rather than purely artistic or scientific, motives for Wood's desire to employ this new labor-saving musical device. His Queen's Hall orchestra had been seriously affected by mass resignations of musicians in May 1904 after he had tried to ban the deputy system. Left with only half his orchestra (those who left were to form the London Symphony Orchestra), the weakened state of his string section had caused ongoing comment in the press.[86] Although supported by Parsons, who had paid the weekly salary of the bass player while the Auxetophone was in use, in return for the daily trials and public exposure, the arrangement was not viable in the long term. Wood's correspondence with Parsons, which continued until 1909, gives a picture of a conductor optimistic about the technology but at war with his string players:

> I have every confidence in the ultimate success, both musically and financially, of your auxetophone, as I am quite sure . . . that you will be able to reinforce five stringed instruments . . . sufficiently to combat the complete wood-wind and brass of the Wagnerian orchestra, and then in small towns they will be able to engage five *good* string players instead of twenty, and they will be able to have a complete wind orchestra which . . . owing to the additional expense, is always curtailed . . . of course orchestral musicians will hate you, but they must not be considered: the artistic result is the only point to keep in view.[87]

Clearly, the rank-and-file orchestral musician objected vigorously to the Auxetophone, as even Parsons himself foresaw a reduction of up to 80 percent in the string sections.[88] To his credit, Wood anticipated not only a rebalancing of orchestral forces, but perhaps even more importantly, he foresaw a radical reordering of the aural landscape of modern orchestral music. One can only speculate that Wood's unflagging enthusiasm, and his assurance of the provision of ready opportunities to test the instruments in concert performance, may well have been a factor in Parsons's continued engagement with the project. In the longer term however, it seemed that the continued resistance by musicians proved to be a major setback in the progress of this emerging technology.

Unbridled industrial growth and technological advances in the late nineteenth and early twentieth centuries had a tumultuous effect on the livelihoods of skilled workers. But the birth of sound recording, the popularity of silent films with musical accompaniment, and other new forms of entertainment led to the creation of thousands of new jobs in what were boom years for musicians. It was also a time of increasing labor militancy in the face of often ruthless and exploitative management, and rank-and-file musicians were to react no differently than their counterparts in other industries.[89] Rather than Luddism, the reaction toward amplification shown by Henry Wood's orchestra may be viewed as a sign of the increasing organization of musicians and their readiness to act in order to protect their employment. It is fitting then, that what may have been the first time in history that a new form of sound technology would trigger something close to an industrial dispute should have occurred within the strictly hierarchical confines of a classical orchestra.

Henry Wood was never again able to employ an Auxetophone, although its development continued for at least two years after that fateful prom season. Horace Short's experimental work at Heaton Works (Parsons's workshop at Newcastle-on-Tyne) included attempts to amplify instruments other than those of the violin family, such as the harp and piano, but it was found to be impossible to locate a single area of their respective sounding boards that would respond to the vibrations of all the strings. Even with as many as four air valves fitted to a harp, the results were not deemed to be successful nor practical enough to warrant further trials on these instruments.[90]

A close inspection of an air valve made by Short for instrument amplification at the Tyne & Wear Discovery Museum reveals a more highly developed design than other existing prototypes or commercially produced air valves found in museum or archive collections elsewhere.[91] Much of the casing is of cast aluminum, while the moveable comb is made from the alloy magnalium, lighter in weight than aluminum, improving the transmission of vibrations and also more resistant to corrosion from air impurities. The spacing of the slots in the comb and fixed grid would vary in an Auxetophone valve depending on its purpose. Parsons wrote of using one-fiftieth of an inch slots for reproducing sounds from a faint phonograph recording to one-quarter of an inch for a double bass. The spacing or "pitch" of the Discovery Museum valve is between these measurements at 1.4 mm. A reconstruction of this air valve was attempted in 1991 by apprentices at the Royal Ordnance Factory at Birtley, Gateshead. Despite copying the artifact in virtually every detail, sadly their tests produced no sound whatsoever (Figure 1.14).[92]

Such a valve would have been mounted on a string instrument by attaching it to a wooden beam that stretched beneath the bridge and clamped to the sides of the instrument's belly on felt pads. A rod transmitting vibrations to the valve was fastened to the

FIGURE 1.14
An Auxetophone air valve designed to amplify a double-bass. Approximately 10.25 × 4.7 × 4.7 inches. Discovery Museum, Tyne & Wear Archives. Photo by Aleksander Kolkowski.

bridge, as has been described above, while long lengths of rubber tubing connected the valve's exit side to a giant coiled horn and its input to a blower placed at distance away to avoid noise interference.[93] This set the blueprint for the way in which modern contact microphones would be applied to acoustic strings (tubing and blower replaced by cables and amplifier), firmly establishing the Auxetophone valve as the world's first instrument transducer pickup or "bug."

A magnificent Parsons-built horn, one of the two used to amplify double basses at the Queen's Hall, stands proudly in the Discovery Museum. Measuring eight feet high, with a bell mouth some three feet wide, the tapering conical horn spirals seventeen feet from a bore of two-and-a-half inches to one-and-a-half feet at its exit. Made of heavy-gauge sheet metal of up to one-quarter-of-an-inch thickness, it does indeed bear more relation to a giant maritime ventilation shaft than to a common gramophone horn. Once again, Parsons foreshadowed the technology of the future as scientific research into the properties of acoustic horns resulted in longer spiraling and folding exponential horns being used decades later with electronic speaker drivers to better reproduce the lower frequency range.[94]

Auguste van Biene and the Auxeto-Cello

In July 1909, the Queen's Hall once again hosted an amplified instrumental performance. This time the instrument was the "auxeto-cello," presented, said the advertisement, for the first time in public.[95] The cellist Auguste van Biene was featured in a program of "modern music de salon," accompanied by the New Symphony Orchestra conducted by Landon Ronald. Van Biene played Boëllmann's *Symphonic Variations*, the slow movement of Lalo's cello concerto in D minor with the orchestra, and a number of solos including a Bach courante and an arrangement of *Ave Maria* (Figure 1.15).[96]

Always a canny businessman, Parsons's choice of van Biene as soloist was a calculated one. The cellist had considerable popular appeal and was well known on the popular, rather than the concert hall, stage. A Dutch-Jewish expatriate who had early in his career busked on the streets of London, van Biene rose to fame through his performances in an internationally successful musical-melodrama, *The Broken Melody*, in which he both acted and played the cello as well as having composed the title tune.[97]

Parsons's decision to hire van Biene may well have been determined by the hostile reaction of professional musicians in Henry Wood's orchestra to assisted amplification. It is quite possible that a more distinguished classical musician may have felt reluctant to promote or play with such an unorthodox attachment to his instrument. Van Biene was certainly somewhat of an outsider as far as the profession was concerned; appearing always to have lived on the edge of respectability,[98] he had worked hard as a performer in popular theater for much of his life and appeared to have few, if any, connections with the elite of the musical establishment. In the music-hall or popular context, the novelty of the auxeto-cello would appear to have been rather more of an advantage than otherwise.

FIGURE 1.15
Auguste van Biene with cello and Auxetophone paraphernalia (ca. 1909). Original caption: *Novo glasbilo za orkester* (New musical instrument for orchestra). Source: *Dom in svet* 22 (1909), National and University Library of Slovenia, Ljubljana (NUK). Reproduced with permission.

A detailed review of Parsons's promotional concert noted that some of the solos were played "on a violoncello to which the auxetophone was attached—an instrument that increases the tone of the violoncello, and was tried for some time by Mr. Wood, who had them attached to some of his double-basses" (Figure 1.16). The review continued:

> Although the volume of tone is undeniably increased, the quality is very much changed, and takes on the brassy quality that one would naturally expect it to do from the brass instruments, a mouthpiece through which the vibrations have to come. This brassy quality is not very unpleasant in itself except when a wide interval is taken, and then the tone seems to vibrate in a harsh way; in any case, the volume is increased at the expense of the characteristic tone of the violoncello.[99]

Such negative reviews and lukewarm appraisals were to sound the death knell for auxeto-assisted instruments. The Queen's Hall was to have been the first of several venues to host van Biene's amplified cello, but no further recitals appear to have taken place.[100]

FIGURE 1.16
View of the Auxetophone air valve attached to cello, showing a connecting rod to the instrument's bridge (ca. 1909). From "Science and Invention," *Literary Digest*, January 1910 (Author's private collection).

FIGURE 1.17
Charles Parsons's Auxetophone horn (ca. 1904), used to amplify the double-bass and gramophone. Featured in a sound installation by the author at the Great North Museum, Newcastle, 2010; loaned by the Tyne & Wear Discovery Museum. Copyright © Louise Hepworth.

Notes

1. From Thomas Edison's invention of the phonograph in 1877 until the introduction of electrical recording in the mid-1920s.

2. The London musical instrument maker A. T. Howson made an early attempt at electrical amplification in 1913. See Howson, Improvements in or Connected with Phonofiddles, Violins, and Other Stringed Musical Instruments. GB Patent 26,143. It provided "a means for electrically transmitting the sound of a musical instrument to a horn or sounding device removed or detached from the instrument itself, the invention being mainly intended for use in connection with the musical instrument known as the phonofiddle, but being also applicable to other stringed instruments." Henri Kubelik, billed as "The Famous Hungarian Eccentric Violin Virtuoso," appears to have played an electrically amplified violin during his 1914 tour of Australia. See "He Would Win Out. The Story of Kubelik, a Talented Violinist," *The Mail*, June 20, 1914, and "Theatre Royal: Henri Kubelik," *The Mercury*, June 5, 1914. Further developments in the area of electrical amplification of string instruments were made by George Beauchamp, who introduced the first versions of his electric violin in 1935. See Richard R. Smith, *Rickenbacker* (Fullerton, CA: Centerstream Publishing, 1987), 53.

3. In addition to archival sources, significant research was conducted at the Tyne & Wear Discovery Museum. A close examination of an Auxetophone valve for instrument amplification designed by Horace Short, and associated contemporary documents has led to a greater understanding of its mode of operation. The examination also shed light on the possible technical reasons for the early demise of this technology. Furthermore, a reconstruction project is planned by Aleks Kolkowski using part of this research as its basis in order to build a working version of the auxeto-cello.

4. The Magnavox moving-coil loudspeaker and public address system was patented in 1913 and first demonstrated in 1915. See Timothy J. Sturgeon, "How Silicon Valley Came to Be," in *Understanding Silicon Valley: The Anatomy of an Entrepreneurial Region*, ed. Martin Kenney (Stanford, CA: Stanford University Press, 2000).

5. Examples of both machines are to be found in the Science Museum collection of communications technology: 1938-313 (Parsons Auxetophone) and 1982-452 (Gaydon Stentorphone).

6. Charles Parsons, Improvements in and Relating to Musical Instruments. GB Patent 10,469.

7. John Matthias Augustus Stroh, Improvements for Violins and Other Stringed Instruments. GB Patent 9418. For detail, see Alison Rabinovici, "Augustus Stroh's Phonographic Violin. A Journey: Victorian London, Australia, Transylvania," *Galpin Society Journal* 58 (May 2005): 100–23.

8. Stoney's letters to Parsons, quoted in extenso, are to be found in Rollo Appleyard, *Charles Parsons: His Life and Work* (London: Constable, 1933), 217–20.

9. Derek B. Scott, *Music, Culture, and Society: A Reader* (Oxford, UK: Oxford University Press, 2000), 129.

10. See, for example, "The Crying Evil," *The Times*, September 19, 1855, 11; "The Polytechnic: Ethiopian Serenaders," *The Times*, May 31, 1858, 12; "Street Musicians," *The Times*, July 2, 1860, 8.

11. Edison, Improvement in Speaking-Machines. US Patent 201,760 (March 4, 1878).

12. "The Aerophone," *New York Times*, March 25, 1878, 4.

13. Quoted in Francis Jehl, *Menlo Park Reminiscences: Part 1* (Dearborn, MI: Edison Institute, 1936), 182. (Jehl was Edison's laboratory assistant.)

14. "A Visit to the Inventor of the Phonograph," *Scientific American Supplement* (April 20, 1878): 1904–05.

15. Edison, Recording and Reproducing Sounds. GB Patent 1644 (April 24, 1878).

16. "The Aerophone."

17. Hughes refused to apply for a patent, preferring to make his discoveries publicly available.

18. "The Language of Flies," *Illustrated Police News*, May 25, 1878.

19. For an account of the history of microphone development, see Bob Paquette, "Early Microphone History," *Antique Wireless Review* (New York, 1989): 131–56.

20. For an overall survey of such machines, see V. K. Chew, *Talking Machines 1877–1914: Some Aspects of the Early History of the Gramaphone* (London: H.M.S.O. Science Museum, 1967), 66–73.

21. Groundbreaking experiments in the recording, reproduction, and transmission of sound with light beams, magnetism, water, and compressed air, had been carried out at Alexander Graham Bell's Volta laboratory during the 1880s. A fine jet of compressed air was used as a stylus to track the grooves on a record in order to reproduce the recorded sounds. This nontactile approach to playing records foresaw laser-scanning techniques almost a hundred years later. See Robert Bruce, *Bell: Alexander Bell and the Conquest of Solitude* (Ithaca, NY: Cornell University Press, 1990), 350–52.

22. A number of these recordings can be accessed at the website of the National Park Service, U.S. Department of the Interior, "Very Early Recorded Sound." http://www.nps.gov/edis/photosmultimedia/very-early-recorded-sound.htm. For the earliest known recorded music in existence, listen to *Israel in Egypt* recorded at the Handel Festival, Crystal Palace, London, in June 1888. Other recordings include the voices of Florence Nightingale, Arthur Sullivan, and Thomas Edison himself.

23. Bruce, "Horace Short, Colonel George E. Gouraud and the Gouraudphone"; H. M. Buist, "Obituary: The Late Mr. Horace L. Short," *Aeronautical Journal* 21 (1917): 456–58.

24. The technology of compressed air, harnessed by French engineers in the 1840s, propelled industrial progress throughout the following decades and was seen by many as a cleaner energy source than electricity. See "Distribution of Power by Compressed Air," *Science* 8, no. 194 (1886): 372–74, and "The Transmission of Energy by Compressed Air," *Science* 14, no. 336 (1889): 29–30. For a comprehensive survey of compressed air technology during the nineteenth century, see John L. Phillips, *The Bends: Compressed Air in the History of Science, Diving, and Engineering* (New Haven, CT: Yale University Press, 1998).

25. Horace L. Short, Improvements in Methods of, and Apparatus for, Increasing the Volume of Sounds from Telephonic or Phonographic Instruments. GB Patent 22,768 (October 29, 1898). Short also registered his patent in Australia in 1899.

26. The Examiner of Patents, Registrar General's Department, Sydney, noted that "the relays and other parts specified are common to telegraphic and telephonic instruments, but the combination, and methods of applying the apparatus, is novel. I recommend that a Patent be granted." See National Archives of Australia, "Correspondence with Horace Leonard Short Concerning Invention Entitled—Improvements in Methods of and Apparatus for Increasing the Volume of Sounds from Telephonic or Phonographic Instruments and Transmitting Such Sounds to Distances." http://naa12.naa.gov.au/SearchNRetrieve/Interface/DetailsReports/ItemDetail.aspx?Barcode=4285902 (accessed May 7, 2010).

27. Horace L. Short, Sound Increasing Device. US Patent 677,476 (July 2, 1901).

28. For an account of the early history of the telephone, see John Brooks, *Telephone: The First Hundred Years* (New York: Harper & Row, 1975).

29. Named by Gouraud after Thomas Edison's research laboratory in Menlo Park, New Jersey.

30. The natural geographical formation of Devil's Dyke, a three-hundred-foot-deep V-shaped valley, may well have assisted the amplification project.

31. "A Shouting Phonograph," *Literary Digest* 21, no. 7 (August 18, 1900): 194.

32. "The Gouraudphone," *Brighton Herald,* September 8, 1900, 7.

33. "The Gouraudphone. Magnifying the Human Voice," *The Star* (New Zealand), December 1, 1900, 1.

34. From Gouraud's announcement in the *Brighton Herald,* August 1, 1900. The Italian tenor Francesco Tamagno was recorded by Short for this purpose and was present during the demonstration. See Bruce, "Horace Short, Colonel George E. Gouraud and the Gouraudphone." Also see "The Gouraudphone: Wonderful Sound Transmitter," *Barrier Miner* (Broken Hill, New South Wales), October 12, 1900, 2. (Quoting from the *London Daily News,* August 28, 1900.)

35. Among the first exhibits were escalators, talking movies, diesel engines, and Campbell's soup.

36. Oswald Short, "Autobiography of the Short Brothers," *Gasbag Magazine, for the Members and Messrs. Short Bros. Cardigan* 1, no. 1 (April 1918): 5. The "liquid air" (most likely composed of nitrogen) was provided by the American scientist and inventor Charles E. Tripler, a fellow exhibitor who took a keen interest in the Gouraudphone. Oswald Short also wrote that the phonograph cylinders reproduced by the "instrument" were of several famous artists, including Francesco Tamagno, who had only days before been recorded by Horace especially for the Eiffel Tower transmission.

37. Charles Cox, "Letter to the Editor," *Hillandale News* 44 (1968): 1271–78.

38. Letter quoted in Gordon Bruce, "Horace Short, The Hon C. A. Parsons and the Auxetophone," *Hillandale News* 97, August 1977.

39. Charles A. Parsons, Improvements in Sound Reproducers or Intensifiers Applicable to Phonographs, Gramophones, Telephones and the Like. GB Patent 10,468 (May 8, 1903), and Improvements in Sound Reproducers or Intensifiers Applicable to Phonographs, Gramophones, Telephones and the Like. GB Patent 20,892 (September 28, 1904).

40. Letter to Sir Ambrose Fleming, 1921, in Appleyard, *Charles Parsons,* 204–05.

41. Gordon Bruce, secretary of Short Brothers, Belfast, writing in *Hillandale News* 97 (August 1977), had uncovered papers at the Science Museum Library detailing correspondence with Short's patent agents Edwards & Co., Parsons's industrial manager S. F. Prest, and Short's own correspondence with Parsons. This paper is indebted to his research.

42. A. Q. Carnegie, "The Parsons Auxetophone," in *Scientific Papers and Addresses of the Hon. Sir Charles A. Parsons, with a Memoir by Lord Rayleigh,* ed. G. L. Parsons (Cambridge, UK: Cambridge University Press, 1934).

43. J. A. Ewing, "The Hon. Sir Charles Parsons, O.M., K.C.B. 1854–1931," *Proceedings of the Royal Society of London: Series A, Containing Papers of a Mathematical and Physical Character* 131, no. 818 (June 1931): vxxv.

44. Parsons, Improvements in Sound Reproducers or Intensifiers Applicable to Phonographs, Gramophones, Telephones and the Like. GB Patent 20,892.

45. Parsons, Improvements in and Relating to Musical Instruments. GB Patent 10,469 (May 8, 1903).

46. Haury, "Claviers à sons Prolongés." *Acoustique et Instruments Anciens: Factures, Musiques et Science,* Cité de la Musique; Société Française d'Acoustique, November 1999, 141–63

47. Parsons, Improvements in and Relating to Musical Instruments. GB Patent 10,469.

48. See Joseph Batten, *Joe Batten's Book: The Story of Sound Recordings* (London: Rockliff, 1956), for photographs and firsthand accounts of the Stroh violin in the recording studio.

49. Stroh had in 1901 patented a new form of diaphragm: Improvement in the Diaphragms of Phonographs, Musical Instruments, and Analogous Sound-Producing, Recording or Transmitting Contrivances. GB Patent 3393. Instead of a flat diaphragm, his was coned with annular corrugations. First used on his violins, this groundbreaking form was later adopted in gramophone soundboxes and went on to influence the design of loudspeakers.

50. For an account of *Turbinia*'s dramatic appearance at Queen Victoria's Jubilee Review, 1897, see Appleyard, *Charles Parsons*, 104.

51. "The Royal Society Conversazione." *The Times*, May 14, 1904, 14.

52. Quoted in Appleyard, *Charles Parsons*, 212.

53. "The 'Auxetophone' at Messrs. Metzler's Rooms," *The Times,* March 21, 1905, 5.

54. Short's first salary payment at the Heaton Works, Tyneside, dates from October 1904. Bruce, "Horace Short, The Hon C. A. Parsons and the Auxetophone."

55. Parsons, Improvements in Sound Reproducers or Intensifiers Applicable to Phonographs, Gramophones, Telephones and the Like. GB Patent 20,892.

56. Parsons, Improvements in and Relating to Sound Reproducers. GB Patent 8407.

57. The London Science Museum has a number of Auxetophone reproducers from 1903 and 1905, with and without the compensating cylinder (1933-357 and 1933-358). The collection also includes a functioning auxeto-gramophone, complete with electric pump and an eight-foot-long horn that was used by Parsons in his demonstrations (1938-313).

58. Appleyard, *Charles Parsons*, 214. This demonstration was the official launch of the commercial Auxetophone by the Gramophone and Typewriter Company.

59. The vertical method of cutting a groove on a cylinder allows for louder sounds to be recorded and reproduced; the dynamic range of early disc records was inherently more limited. Sound quality on a spinning disc is also affected by constant angular velocity (CAV); as the needle reaches the center, the linear distance per revolution decreases, causing sound distortion.

60. Short's silver cylinder is reported in "A Shouting Phonograph." Henri Lioret, in 1897, had demonstrated a celluloid cylinder that gave remarkably loud results.

61. T. R. "The Work of the Late Hon. Sir Charles A. Parsons, O.M., K.C.B., F.R.S.-(5) The Auxetophone," *Heaton Works Journal* (December 1934), though Gordon Bruce states the amount was £2,000.

62. "A Shouting Phonograph."

63. According to the U.S. Census Bureau, the average wage in 1906 for a "blue collar" worker in the United States was between $200 and $400 per annum.

64. Victor Aux-e-to-phone advertisement, *Redbook* magazine. In Britain, the Auxeto-Gramophone was made to order only (according to a 1912 Gramophone Company, Ltd., advertising brochure). Two models existed, a large "concert" auxeto-gramophone and a "home" version, as reported in "Amusements. Theatre Royal. Auxeto-Gramophone," *The Mercury* (Hobart, Tasmania), February 24, 1909, 3.

65. "The Auxetophone," *Illustrated London News*, March 25, 1905, 410.

66. "Times Will Flash the Result." *New York Times*, November 4, 1906, 2.

67. Two such postcards showing thousands of spectators listening to the Auxetophone at Battersea Park and Peckham Rye Common, London, both from 1910, are in the Science Museum Collection. Such open-air demonstrations were widely reported, also, in the Australian press. The *Brisbane Courier* commented in December 1908 that "an Auxeto concert alternated with band

concerts would go a long way towards solving the problem of drawing people from the streets and popularizing the Gardens as a place of resort on summer nights."

68. "Direct Comparison" [Letter to the Editor], *The Gramophone*, April 1933, 40.

69. "The Auxeto-Gramophone Recital," *Clarence and Richmond Examiner* (Grafton, New South Wales), November 21, 1908, 5. One of many contemporary reports in the Antipodean press of Auxeto-Gramophone recitals during 1908 and 1909, following a five-month promotional tour under the direction of S. H. Sheard, of the Gramophone Company, Ltd. Some mention images of the featured artists "thrown" on a large screen.

70. *Hillandale News* 44 (August 1968).

71. "A Gramophone Concert." *The Times,* March 2, 1909, 12.

72. "Direct Comparison."

73. For examples, see "Auxetophone Orchestrations," http://www.auxetophone.com/Orchestrations.html.

74. Daily Auxetophone concerts with a live band also took place at Madame Tussaud's in London from 1908. See "Madame Tussaud's," *The Times*, February 14, 1908, 13.

75. Oliver Read and Walter L. Welch, *From Tin Foil to Stereo: Evolution of the Phonograph* (Indianapolis, IN: Howard W. Sams, 1959), 202–04.

76. "Madame Tussaud's." This review was typical of many others at the time.

77. Emily Thompson argues that the Edison Tone Tests (April 1916) were the first step in the public acceptance of listening to reproduction as equivalent to listening to live performers, whereas the process had already begun with the Auxetophone a decade earlier. See Emily Thompson, *The Soundscape of Modernity* (Cambridge, MA: MIT Press, 2002), 237–38.

78. "The Auxetophone."

79. Appleyard, *Charles Parsons*, 212–13.

80. Henry Wood's youthful interest in engineering and technology suggests that he was well able to understand and appreciate both the function and the potential of Parsons's invention. See Henry J. Wood, *My Life of Music* (London: Purnell & Sons, Ltd.), 15–32.

81. Gamba, "Violinists at Home and Abroad," *The Strad*, August 1906, 114.

82. Carnegie, "The Parsons Auxetophone."

83. Gamba, "Violinists at Home and Abroad," 222.

84. Although it is quite clear that only one bass was amplified, the presence of two horns suggests that perhaps a second amplified bass was envisaged.

85. Carnegie, "The Parsons Auxetophone."

86. "Sir Henry J. Wood," and Gamba, "Violinists at Home and Abroad," 164.

87. Letter to Charles Parsons, June 30, 1909, in Carnegie, "The Parsons Auxetophone," 215–16.

88. Carnegie, "The Parsons Auxetophone," 245.

89. See James P. Kraft, *Stage to Studio: Musicians and the Sound Revolution, 1890–1950*, Studies in Industry and Society 9 (Baltimore, MD: The Johns Hopkins University Press, 1996), 22–23.

90. Carnegie, "The Parsons Auxetophone."

91. From a comparison with valves examined at the Science Museum's Blythe House, West London; Discovery and Beamish Museums, Northeast England, and the EMI Archives at Hayes, Middlesex.

92. A. Howe, D. Oldham, and M. Weightman, "Report by Students Undertaking a BTEC HNC Course in Mechanical Engineering, on a Final-Year Project Reconstructing an Auxetophone Value," Archives, Tyne & Wear Discovery Museum.

93. Carnegie, "The Parsons Auxetophone"; "One Instrument Played by Another," *The Literary Digest*, January 1910, 2, 95; "Catalogue of Parsons Exhibition, North East Coast Institution of Engineers & Shipbuilders," November 1936.

94. John Liffen describes such a horn in detail: "The 1930 Demonstration Broadcast Receiver and Loudspeaker at the Science Museum, London," *British Vintage Wireless Society* Bulletin 31, no. 4 (2006). A. Q. Carnegie wrote that Parsons himself tested the Auxetophone valve using a Western Electric logarithmic horn in 1930 with good results.

95. "Van Biene of 'Broken Melody' Fame," *The Times,* July 7, 1909, 8.

96. "Sir Henry J. Wood," and Gamba,"Violinists at Home and Abroad," 164.

97. For a more detailed account of van Biene, see George Kennaway, "The Phenomenon of the Cellist Auguste van Biene: From the Charing Cross Road to Brighton via Broadway," *Victorian Soundscapes Revisted*, ed. Martin Hewitt and Rachel Cowgill (Horsforth, Leeds Trinity, and All Saints/Leeds Centre for Victorian Studies, 2007); Brenda Neece, "Magician of the Cello. Auguste van Biene's Brilliant Career," *The Strad* 112, no. 1338 (2001): 1102–09.

98. Van Biene appeared in court on a number of occasions for failing to pay the members of his company. See for instance, "The Broken Melody. Van Biene Arrested. An Unsatisfied Debt," *Adelaide Advertiser*, August 14, 1905. For court appearances for breach of copyright see, for instance, "'Maritana' in Court," *The Era*, March 19, 1881, 5.

99. "Mr. Auguste van Biene's Recital," *The Times*, July 9, 1909, 13.

100. Parsons's correspondence concerning the planned auxeto-concert series is referred to in W. Garrett Scaife, *From Galaxies to Turbines: Science, Technology, and the Parsons Family* (Philadelphia: Institute of Physics Publishing, 2000), 471–520.

101. Letter to Sir Ambrose Fleming, 1921, in Appleyard, *Charles Parsons*, 205.

102. Bruce, "Horace Short, The Hon C. A. Parsons and the Auxetophone."

103. Christopher Henry Barnes, *Short's Aircraft since 1900* (London: Putnam Aeronautical), 1989, 47.

104. Gamba, "Violinists at Home and Abroad,"114.

105. "Instructions for the Care and Operation of the Victor Auxetophone," Owner's Manual, Victor Talking Machine Company, Camden, NJ: circa 1907.

106. From the recollections of Richard Cole, former Science Museum curator and Auxetophone operator. A note found in the Science Museum Document Centre (T/1938-313) even recommends the best disc to be played: H.M.V. C7745, Casse-Noisette Suite, op. 71a (Tchaikovsky): Marche, recorded 1917.

107. Bruce, "Horace Short, The Hon C. A. Parsons and the Auxetophone."

108. Amanda S. Barnard, *The Diamond Formula: Diamond Synthesis—A Gemmological Perspective* (Butterworth-Heinemann, 2010).

109. An account of staff member A. Q. Carnegie's experiments is found in T. R., "The Work of the Late Hon. Sir Charles A. Parsons, O.M., K.C.B., F.R.S.-(5) The Auxetophone."

110. Herbert Jüttemann, *Phonographen und Grammophone* (Braunschweig, Germany: Klinkhardt & Biermann, 1979), 229–32.

111. Georges Mendel, Improvements in or Relating to Phonograph and Like Apparatus. GB Patent 8808.

112. The Gaumont Chronophone is pictured by Douglas Self in his comprehensive Auxetophone website pages: "The Short but Loud Story of Compressed-Air Amplification." http://www.aqpl43.dsl.pipex.com/MUSEUM/COMMS/auxetophone/auxetoph.htm#gau (accessed July 3, 2001). Examples of specially pressed records for films that were pneumatically reproduced, by

Gaumont and others, are to be found on the website of Henri Chamoux, "Quelques disques de synchronisation pour le cinéma." http://www.archeophone.org/cinema/ (accessed October 1, 2010).

113. Steven Halliday, "A Pressing Problem—A History of Congestion in London's Tube," *Berkhamsted Review*, November 2001. The Stentorphone was first used on the London Underground at Charing Cross and Oxford Circus stations in January 1921. A cartoon lampooning its use was published in *The Star*, January 28, 1921.

114. Bruce, "Horace Short, The Hon C. A. Parsons and the Auxetophone."

115. Letter to Sir Ambrose Fleming, 1921, in Appleyard, *Charles Parsons*, 20.

Bibliography

"The Aerophone." *The New York Times*, March 25, 1878, 4.

"Amusements. Theatre Royal. Auxeto-Gramophone," *The Mercury* (Hobart, Tasmania), February 24, 1909, 3.

Appleyard, Rollo. *Charles Parsons: His Life and Work*. London: Constable, 1933.

"The Auxeto-Gramophone Recital." *Clarence and Richmond Examiner* (Grafton, New South Wales), November 21, 1908, 5.

"Auxeto-Gramophone Recital." *Brisbane Courier*, December 14, 1908, 6.

"The Auxetophone." *Illustrated London News*, March 25, 1905, 410.

"The 'Auxetophone' at Messrs. Metzler's Rooms." *The Times*, March 21, 1905, 5.

"Auxetophone Orchestrations." http://www.auxetophone.com/Orchestrations.html (accessed October 28, 2007).

Barnard, Amanda S. *The Diamond Formula: Diamond Synthesis—a Gemmological Perspective*. London: Butterworth-Heinemann, 2000.

Barnes, Christopher Henry. *Short's Aircraft since 1900*. London: Putnam Aeronautical, 1989.

Batten, Joseph. *Joe Batten's Book: The Story of Sound Recording*. London: Rockliff, 1956.

"The Broken Melody. Van Biene Arrested. An Unsatisfied Debt." *Adelaide Advertiser*, August 14, 1905.

Brooks, John. *Telephone: The First Hundred Years*. New York: Harper & Row, 1975.

Bruce, Gordon. "Horace Short, Colonel George E. Gouraud and the Gouraudphone." *Hillandale News* 96, June 1977.

———. "Horace Short, The Hon C. A. Parsons and the Auxetophone." *Hillandale News* 97, August 1977.

Bruce, Robert V. *Bell: Alexander Bell and the Conquest of Solitude*. Ithaca, NY: Cornell University Press, 1990.

Buist, H. M. "Obituary: The Late Mr. Horace L. Short." *Aeronautical Journal* 21 (1917): 456–458.

Carnegie, A. Q. "The Parsons Auxetophone." In *Scientific Papers and Addresses of the Hon. Sir Charles A. Parsons, with a Memoir by Lord Rayleigh*, ed. G. L. Parsons. Cambridge, UK: Cambridge University Press, 1934.

"Catalogue of Parsons Exhibition, North East Coast Institution of Engineers & Shipbuilders," November 1936.

Chamoux, Henri. "Quelques disques de synchronisation pour le cinema." http://www.archeophone.org/cinema/ (accessed October 1, 2010).

Chew, V. K. *Talking Machines 1877–1914: Some Aspects of the Early History of the Gramophone*. London: H.M.S.O. Science Museum, 1967.

Cox, Charles. "Letter to the Editor." *Hillandale News* 44 (1968): 1271–78.
"The Crying Evil." *The Times*, September 19, 1855, 11.
"Direct Comparison" [Letter to the Editor], *The Gramophone*, April 1933, 40.
"Distribution of Power by Compressed Air." *Science* 8, no. 194 (1886): 372–74.
Edison, Thomas A. Improvement in Speaking-Machines. US Patent 201,760 (March 4, 1878).
———. Recording and Reproducing Sounds. GB Patent 1644 (April 24, 1878).
Ewing, J. A. "The Hon. Sir Charles Parsons, O.M., K.C.B. 1854–1931." *Proceedings of the Royal Society of London: Series A, Containing Papers of a Mathematical and Physical Character* 131, no. 818 (June 1931): vxxv.
Gamba, "Violinists at Home and Abroad." *The Strad*, August 1906, 114, 222.
"The Gouraudphone." *Brighton Herald,* September 8, 1900, 7.
"The Gouraudphone. Magnifying the Human Voice." *The Star* (New Zealand), December 1, 1900, 1.
"The Gouraudphone: Wonderful Sound Transmitter." *Barrier Miner* (Broken Hill, New South Wales), October 12, 1900, 2.
"A Gramophone Concert." *The Times,* March 2, 1909, 12.
Halliday, Steven. "A Pressing Problem—A History of Congestion on London's Tube." *Berkhamsted Review*, November 2001.
Haury, Jean. "Claviers à sons Prolongés." *Acoustique et Instruments Anciens: Factures, Musiques et Science*, Cité de la Musique; Société Française d'Acoustique, November 1999, 141–63.
"He Would Win Out. The Story of Kubelik, a Talented Violinist." *The Mail*, June 20, 1914, 10.
Hillandale News 44, August 1968.
Howe, A., D. Oldham, and M. Weightman. "Report by Students Undertaking a BTEC HNC Course in Mechanical Engineering, on a Final-Year Project Reconstructing an Auxetophone Valve." Tyne & Wear Museum, Newcastle.
Howson, Arthur Thomas. Improvements in or Connected with Phonofiddles, Violins and Other Stringed Musical Instruments. GB Patent 26,143 (November 14, 1913).
"Instructions for the Care and Operation of the Victor Auxetophone." Owner's Manual, Victor Talking Machine Company. Camden, NJ: circa 1907.
Jehl, Francis. *Menlo Park Reminiscences: Part 1*. Dearborn, MI: Edison Institute, 1936.
Jüttemann, Herbert. *Phonographen und Grammophone.* Braunschweig, Germany: Klinkhardt & Biermann, 1979, 229–32.
Kennaway, George. "The Phenomenon of the Cellist Auguste van Biene: From the Charing Cross Road to Brighton Via Broadway." In *Victorian Soundscapes Revisited*, ed. Martin Hewitt and Rachel Cowgill. Horsforth, Leeds Trinity and All Saints/Leeds Centre for Victorian Studies, 2007, 67–82.
Kraft, James P. *Stage to Studio: Musicians and the Sound Revolution, 1890–1950.* Studies in Industry and Society, 9. Baltimore, MD: The Johns Hopkins University Press, 1996.
Langley, Leanne. "Building an Orchestra, Creating an Audience: Robert Newman and the Queen's Hall Promenade Concerts, 1895–1926." In *The Proms: A New History*, ed. Jenny Doctor, David Wright, and Nicholas Kenyon. London: Thames and Hudson, 2007, 32–73.
"The Language of Flies." *Illustrated Police News*, May 25, 1878, 4.
Liffen, John. "The 1930 Demonstration Broadcast Receiver and Loudspeaker at the Science Museum, London." *British Vintage Wireless Society Bulletin* 31, no. 4, 2006.
"Madame Tussaud's." *The Times*, February 14, 1908, 13.
"'Maritana' in Court." *The Era*, March 19, 1881, 5.

Mendel, Georges. Improvements in or Relating to Phonograph and Like Apparatus. GB Patent 8808 (August 8, 1907).

"Mr. Auguste van Biene's Recital." *The Times*, July 9, 1909, 13.

National Archives of Australia. "Correspondence with Horace Leonard Short Concerning Invention Entitled—Improvements in Methods of and Apparatus for Increasing the Volume of Sounds from Telephonic or Phonographic Instruments and Transmitting Such Sounds to Distances." http://naa12.naa.gov.au/SearchNRetrieve/Interface/DetailsReports/ItemDetail.aspx?Barcode=4285902 (accessed May 7, 2010).

National Park Service, U.S. Department of the Interior, "Very Early Recorded Sound." http://www.nps.gov/edis/photosmultimedia/very-early-recorded-sound.htm (accessed August 1, 2011).

Neece, Brenda. "Magician of the Cello. Auguste van Biene's Brilliant Career." *The Strad* 112, no. 1338 (2001): 1102–09.

"One Instrument Played by Another." *The Literary Digest*, January 1910, 2, 95.

Paquette, Bob. "Early Microphone History." *The Antique Wireless Review* (New York 1989): 131–56.

Parsons, Charles A. Improvements in and Relating to Musical Instruments. GB Patent 10,469 (May 8, 1903).

———. Improvements in and Relating to Sound Reproducers. GB Patent 8407 (April 19, 1905).

———. Improvements in Sound Reproducers or Intensifiers Applicable to Phonographs, Gramophones, Telephones and the Like. GB Patent 10,468 (May 8, 1903).

———. Improvements in Sound Reproducers or Intensifiers Applicable to Phonographs, Gramophones, Telephones and the Like. GB Patent 20,892 (September 28, 1904).

Phillips, John L. *The Bends: Compressed Air in the History of Science, Diving, and Engineering*. New Haven, CT: Yale University Press, 1998.

"The Polytechnic: Ethiopian Serenaders." *The Times*, May 31, 1858, 12.

Rabinovici, Alison. "Augustus Stroh's Phonographic Violin. A Journey: Victorian London, Australia, Transylvania." *Galpin Society Journal* 58 (May 2005): 100–23.

Read, Oliver, and Walter L. Welch. *From Tin Foil to Stereo: Evolution of the Phonograph*. Indianapolis, IN: Howard W. Sams, 1959.

"The Royal Society Conversazione." *The Times*, May 14, 1904, 14.

Scaife, W. Garrett. *From Galaxies to Turbines: Science, Technology, and the Parsons Family*. Philadelphia: Institute of Physics Publishing, 2000.

Scott, Derek B. *Music, Culture, and Society: A Reader*. Oxford, UK: Oxford University Press, 2000.

Self, Douglas. "The Short but Loud Story of Compressed Air Amplification." http://www.aqpl43.dsl.pipex.com/MUSEUM/COMMS/auxetophone/auxetoph.htm#gau (accessed July 3, 2001).

Short, Horace L. Improvements in Methods of, and Apparatus for, Increasing the Volume of Sounds from Telephonic or Phonographic Instruments, and Transmitting Such Sounds to Distances. GB Patent 22,768 (October 29, 1898).

———. Sound Increasing Device. US Patent 677,476 (July 2, 1901).

Short, Oswald. "Autobiography of the Short Brothers." *Gasbag Magazine, for the Members and Messrs. Short Bros. Cardigan* 1, no. 1 (April 1918): 5.

"A Shouting Phonograph." *Literary Digest* 21, no. 7 (August 18, 1900): 194.

"Sir Henry J. Wood." *The Musical Times* 52, no. 817 (March 1, 1911): 156.

Smith, Richard R. *Rickenbacker*. Fullerton, CA: Centerstream Publishing, 1987.

"Street Musicians." *The Times*, July 2, 1860, 8.

Stroh, John Matthias Augustus. Improvement in the Diaphragms of Phonographs, Musical Instruments, and Analogous Sound-Producing, Recording or Transmitting Contrivances. GB Patent 3393 (December 14, 1901).

———. Improvements for Violins and Other Stringed Instruments. GB Patent 9418 (May 4, 1899).

Sturgeon, Timothy J. "How Silicon Valley Came to Be." In *Understanding Silicon Valley: The Anatomy of an Entrepreneurial Region*, ed. Martin Kenney. Stanford, CA: Stanford University Press, 2000, 30–31.

"Theatre Royal Henri Kubelik." *The Mercury*, June 5, 1914, 3.

Thompson, Emily. *The Soundscape of Modernity*. Cambridge, MA: MIT Press, 2002.

"Times Will Flash the Result." *New York Times*, November 4, 1906, 2.

T. R. "The Work of the Late Hon. Sir Charles A. Parsons, O.M., K.C.B., F.R.S.-(5) The Auxetophone." *Heaton Works Journal* (December 1934): 257.

"The Transmission of Energy by Compressed Air." *Science* 14, no. 336 (1889): 29–30.

"Van Biene of 'Broken Melody' Fame," *The Times,* July 7, 1909, 8.

Victor Aux-e-to-phone advertisement, *Redbook* magazine, June 1907.

"A Visit to the Inventor of the Phonograph." *Scientific American Supplement* (April 20, 1878): 1904–05.

Wood, Henry J. *My Life of Music*. London: Purnell & Sons, Ltd., 1938.

CHAPTER 2

Artifacts in Performance

Katy Price

Introduction

When artifacts are used in performance, the audience may experience technology through music in ways that heighten awareness of the past while at the same time sparking reflection on cultural production and consumption in the present. Showmanship and live curating can open up the history of technology to fresh audiences (or to familiar ones in new ways). This combination of education with entertainment also has the potential to affect how audience members perceive everyday relationships with technology through work and leisure. Viewers and listeners can never be forced to think or feel along predetermined lines, but any text or performance may be analyzed to see what opportunities are presented for active engagement with aesthetic or cultural questions.

This chapter analyzes two examples of artifacts in performance: the ensemble Apartment House performing a Stroh quartet at Kettle's Yard, Cambridge (March 2006), and a concerto for iPhone and Pianola by Julio d'Escriván with Rex Lawson at West Road Concert Hall, Cambridge (November 2009). The discussion blends technical detail with consideration of audience experience. Information that might be considered incidental from a strictly musical point of view, such as a pre-performance talk or an adjustment during the concert, is drawn in where relevant. Before turning to the examples, I explore briefly the question of effort in relation to music technology and audience experience, following a recent discussion by d'Escriván himself, "To Sing the Body Electric: Instruments and Effort in the Performance of Electronic Music" (2006). Drawing theoretical perspectives together with historical detail and performance analysis, I seek to demonstrate how artifacts can be used to enrich audience awareness of present conditions through a dynamic representation of past episodes.

A Stroh instrument is a violin or any other stringed instrument in which the wooden resonating body has been replaced by a diaphragm and metal horn. A Pianola is a piano fitted with a mechanism for playing the notes, activated by a punched roll. These two examples have been deliberately chosen as coming from the late 1800s to early 1900s, a period that pre-dates electronic music but nevertheless saw experiments in the production and reproduction of sound. The intention here is to illuminate a central concern of the *Artefacts: Music* volume, which asks whether "the application of electronics marks a distinct phase in the development of music, or whether the application of new technologies has merely exaggerated existing tendencies." As Hans-Joachim Braun, introducing *Music and Technology in the Twentieth Century* (2002), observes, machines entered music long before electronics.[1] Has the transition from mechanical to electromechanical to electronic instruments really been as radical as we might like to imagine, from a twenty-first-century perspective in which life without microprocessing seems primitive? Or conversely, have the changes been more radical than we are habitually able to conceive of, having grown so used to present modes of interaction through electronic devices?

The 1880s to 1920s constitute a peculiar period in the history of technology, at once marching relentlessly toward the future and remaining resolutely Victorian. Edison's 1877 phonograph, a purely mechanical means of recording and reproducing sound, may be viewed as a humble precursor to developments in electronic, amplified sound, or as the triumph of mechanical understanding and control of sound. Stroh instruments and player pianos emerged during the 1890s and had declined in use by the end of the 1930s, largely thanks to the rise of electronic amplification.[2] As mechanical innovations, they first benefited from and were then rendered obsolete by developing markets for recording technology, occupying an ambiguous cultural niche that is explored in more detail below. In the context of early twenty-first-century performance, such ambiguities begin to make trouble for a straightforward choice between electronics as "a distinct phase in the development of music" or as an exaggeration of "existing tendencies." The distinctive quality of these instruments in performance is brought out through a discussion of "effort" as a contentious theme in electronic (as opposed to acoustic and electroacoustic) music.

Instruments and Effort

D'Escriván's essay on "Instruments and Effort" addresses the question of whether electronic music can be truly satisfying for a concert audience. The advent of microprocessing raises two key issues: loss of visible technique or skill, and reduced physical exertion. Where early electronic instruments such as the Theremin or Ondes Martenot require practiced manipulation of gestures that bear a recognizable relation to output sound, devices used with computers (such as game controllers, gloves, or helmets) may produce sound in any way the programmer chooses, often removing physical effort altogether.[3] In the context of music with computers, physical symptoms such as perspiration have come to signify a generation gap: musicians "under, say, thirty years of age" are "not accustomed to seeing much sweat in the performance of electronic music."[4]

A composer and performer of electronic music himself, d'Escriván asks how far "performing effort" and "bodily involvement" are necessary for an audience to be engaged.[5] "How much control should the performer have on the sound for a performance to be exciting? . . . Does it also matter if the objects that we designate as instruments do not look like they could even possibly correlate with their assigned sonic output?"[6] He concludes that the answer may vary according to age and experience: "Those who have been brought up with personal computers and video games could be more open towards *effortless* performances. People of an older generation may tend to require an old-school paradigm of performing virtuosity, where perceived effort and dexterity on behalf of the performer are paramount to the enjoyment of music."[7] The prominence of this theme in writing and conversation about contemporary music suggests that the question of physical effort is attended by cultural concerns in addition to the generational tension.

Ideologies of Sound Production

Ambivalence about physical presence is a key theme of David Toop's *Haunted Weather* (2004). "As the apparatus of music becomes less apparent, particularly in the digital domain," he observes, "so sound becomes more completely itself, the purest manifestation of a disembodied, time based art."[8] Toop includes a broad sweep of popular and classical forms in opposition to the laptop: "Freed from the distraction of ranked violinists dressed in black and white sawing at their instruments, guitarists leaping around on a stage, entire typing pools of keyboard players, choreographed dancers, drum risers, video walls and pyrotechnics, the intangible core of music, the part that makes some people close their eyes when they listen, is allowed its full power." But, he acknowledges, "This gain comes with some profound losses. Whether based on false assumptions or deep-seated needs, the sight of musicians playing in real time, engaged in actions that have a discernible link to the sounds they are producing, makes an audience feel a warm glow of communication."

Paul Théberge has also observed that the "direct relationship between physical gestures and sound" is "completely severed with electronic devices," giving sounds "apparent autonomy and uncommon power [. . .] in determining how you play them."[9] His argument in *Any Sound You Can Imagine* (1997) pursues the broader social and industrial context for this shift in musical ethos, capturing a deep tension between "romantic ideologies of personal expression that have been traditionally associated with musical instruments" and the more recently evolved condition of musicians as producer-consumers.[10] The loss of a physical relationship to sound through MIDI, computers, and reproduction technology, along with fears about the departure of skill and creativity, is part of a much broader shift from Victorian industrial society to "postmodern capitalist enterprise."[11] In this context, the very meaning of "sound" and "live" have shifted but still retain something of their romantic associations and values.[12]

Théberge's tension between romantic and late capitalist ideologies of sound production offers a broad conceptual framework through which we may appreciate different arguments about sound generation and the performing body. This is useful because passions can run high on the topic. Toop's pure sound versus social act is expressed in a measured, almost detached temper, though he hankers to educate audiences in the appreciation of pure sound and attaches a hint of frustration to the continuing need for that saccharine "warm glow." Opinions during the 1970s were articulated in stronger terms. Murray Schafer, in his 1977 account of soundscape transformation through the industrial and electric revolutions, identifies "schizophonia" as a problematic twentieth-century development. The term is defined as an act of violence, with uncanny effects: "We have split the sound from the maker of the sound. Sounds have been torn from their natural sockets and given an amplified and independent existence. Vocal sound, for instance, is no longer tied to a hole in the head but is free to issue from anywhere in the landscape."[13]

Roland Barthes, in "The Grain of the Voice" (1972), adapts from Julia Kristeva a distinction between "phenotext" and "genotext," drawing out two opposed modes of singing. The "phenosong," which Barthes found ubiquitous, constitutes "everything in the performance which is in the service of communication, representation, expression."[14] This he associated with facile notions of embodiment, a "myth of respiration," which he dismissed with a visceral image: "The lung, a stupid organ (lights for cats!)."[15] The phenosong dominates through mass culture, with its concern for sentiment, drama, and clear expression. In contrast, the "genosong" bears a deeper relation to both language and the body, being "the volume of the singing and speaking voice," and "having nothing to do with communication, representation (of feelings), expression."[16] The histrionic energy directed by Barthes against the phenosong, which lends itself to elaboration much more fully than does the mysterious "grain" of his essay's title, may possibly function as a parody of that category's action. Even so, it gives the essay itself a distinctly "phenotextual" edge.

Schafer and Barthes were writing well before the adoption of game controllers in electronic music, and only just contemporaneously with the emergence of microprocessing. Their rhetoric serves as a reminder that there is more at stake in the contested relationship between physical gesture and output sound than simply the accommodation or rejection of computer technology in music.

A Theory of Sentics

D'Escriván pursues an analogy with space exploration as a provisional answer to the problem of communication in musical performance lacking "sweat." This is borrowed from a "theory of sentics" developed in 1970 by Manfred Clynes, in the context of concerns about mental health on space missions.[17] Clynes explored the capacity for minimal physical actions to become associated with emotional expression, thereby offering astronauts a way to experience a rich, affective life in cramped conditions. The subtitle of an article by Clynes on his research in *Psychology Today* (1972), "Sentic Cycles: The Passions at Your Fingertips," captures the general idea. D'Escriván suggests that the attachment of emotional intent to the click of a mouse may give performance with computers a "sentic" significance. As he observes, the key difference between an astronaut expressing feelings and a musician performing is the absence or presence of an audience. The space traveler knows what he or she intends. But how will the laptop artist's listeners know what emotions have been intended by each click?

Minimal gestures are nothing new in Western art or popular music. In the acousmatic tradition of musique concrète, gesture is absent altogether, while John Cage's 4'33" makes a performance out of the refusal of gesture, throwing intention open to question. Electronic dance music covers the full spectrum, from the "sentics" of Kraftwerk (as one YouTube comment put it, "A bit dull live arent they? I wouldent pay £40 to see 4 balding germans press a few keys on a laptop") to the histrionics of Throbbing Gristle. The gestures available to a performer are not simply determined by technology: expectations about the necessity (and form) of spectacle also play a role. Trevor Pinch and Frank Trocco, writing about early Moog marketing, illustrate the point nicely. Analyzing a promotional photograph of the Series 900 modular synthesizer, they note "a posture deliberately used by Moog in his advertising," in which the musicians at work have their left hands adjusting knobs while the right hand is lower, preferably resting on a keyboard.[18] This gesture "seemed to encapsulate the link between the music and the machine," and Moog was clearly aware of the need to meet existing expectations about the generation of sound in music: "It looks good if you're playing a keyboard," he remarked. "People understand that then you're making music. You know [without it] you could be tuning in Russia!"

The radio joke shows awareness of audience unease in relation to unfamiliar modes of performance. Without a keyboard, Moog operation is at once inaccessible and subversive (tuning in to Russia), and, paradoxically, all too accessible (anyone might do it by accident, so where is the skill?). The gestures associated with new technology do trigger anxieties about accessibility and skill, but these extend beyond the kit and its uses. As Théberge insists, "How you learn to make and listen to music cannot be explained solely by the direct physical or cognitive relationship between you and your chosen instrument."[19] Music theorists, he observes, tend to focus their concerns about "the creative role of new technology" on "problems of human/machine 'interaction,'" while "problems of a more collective or social nature" slide out of view. In the two performances analyzed below, the use of artifacts offers the audience an opportunity to engage with innovations in sound production in ways that invoke a social context. The Stroh instruments and Pianola, with their ambiguous position in the history of recording technology, drive a performance aesthetic that is "sentic" and "histrionic" by turns, prompting self-consciousness about music technology in tension between romantic and late capitalist ideologies of musical production.

Stroh Instruments

Stroh instruments are strings (notably violins, but also viola, cello, bass, guitar, mandolin, ukulele, and one-string fiddle) on which the wooden sounding body has been replaced by a diaphragm and horn, enabling a more strident and directional sound. They were manufactured in London from 1904 to 1942, with various spin-offs and adaptations occurring under other names around the world.[20] The Stroh carries the aura of "a quaint oddity from a bygone age, a period notable for inventions, often more bizarre than practical."[21] Yet they were invented by a man of great mechanical acumen, and they were used almost universally in early recording studios, replacing conventional stringed instruments as these were not captured well before the advent of microphones and electric amplification. Rendered obsolete by those developments in the mid-1920s, Strohs continued to be featured in dance bands, Morris dancing, and the music hall, and may occasionally be heard today among buskers and in folk music.

Augustus Stroh (1828–1914) was an exemplary modern Victorian, setting out as a watchmaker and moving on to work at the forefront of telegraphy, sound recording, and acoustics.[22] When news of Edison's phonograph reached British audiences in 1878, Stroh was commissioned to build a demonstration model, to which he added certain improvements.[23] His phonographic violin, incorporating the diaphragm and horn from Edison's machine, was patented in 1899 (Figure 2.1). Two years later Stroh made the crucial adaptation of a cone-shaped diaphragm (enabling uniform vibrations across its surface), an innovation with "far-reaching consequences, not only for diaphragms used by gramophone, phonographs and Stroh violins, but also for the development of electric loudspeaker design some twenty-five years later."[24] Alison Rabinovici notes the existence of plans for a musical instrument incorporating a diaphragm among Stroh's acoustical experiments for 1879, casting received mythology about the invention into doubt: "Contrary to the popular view; Stroh did not develop his amplified violin purely as a response to a practical need for a louder and more directional violin for an as yet imperfect recording technology."[25] The Stroh violin, she suggests, grew out of explorations stimulated by the phonograph, but it was not simply devised to address a comparative weakness in conventional stringed instruments. The Stroviols trademark was registered in 1910 by George Evans, who added a secondary horn (enabling the player to monitor sound) during the 1920s.[26]

The Stroh name, adopted in Evans's trademark, was a marker of distance from more populist musical innovations. Rabinovici elaborates on the distinction between Stroh instruments and the phonofiddle, a single-stringed, horned violin. This trademark was registered in 1906 by the inventor A. T. Howson, who "supplied the music hall entertainers with 'curiosity' instruments that were used effectively in comedy routines and, more importantly, provided much needed amplification in large and crowded music halls."[27] The phonofiddle had a ready route into the music hall following its nonhorned predecessor, the so-called Japanese or Jap fiddle, and was taken up by proponents of the

FIGURE 2.1
A recording session with Rosario Bourdon conducting the Victor Salon Orchestra. A Stroh violin is visible in the foreground. Courtesy of the George H. Clark Radioana Collection, Archives Center, National Museum of American History, Smithsonian Institution.

earlier instrument, most famously George Chirgwin. While Stroh instruments designed for recording studio use "required trained musicians to play them," a "performance on the single stringed phonofiddle was much more within the reach of the musician who lacked formal or extensive classical training." Promotional materials and press notices for the phonofiddle stressed ease of playing and accessibility to beginners. "It is almost as easy to play the Phonofiddle by ear as it is to whistle or hum a tune," Laura Howson (the inventor's spouse) advised in her 1910 tutor for the instrument.[28] For all its associations with the scientific and classical music establishment, a similar vein emerged in Stroh coverage: "Although the diaphragm is made of metal aluminum there is no metallic sound audible. . . . The rich mellow tones . . . require no forcing. The slightest contact of the bow will bring them forth, and make the player imagine himself a far better player than he really is."[29] Such rhetoric of enhanced performance would already have been familiar thanks to its widespread use in automatic piano marketing from the late 1890s onward.[30]

An Aesthetic of Histrionics

The character of Stroh sound, like any auditory experience, is shaped to an extent by the listener's expectations. Contemporary accounts tended to note a significant improvement on regular stringed instruments sound. "The G string is a dream. It possesses the deep rich quality of a fine 'cello A, but there is no unevenness in the strings. The harmonics are loud and pure, and what is of great importance is an entire absence of 'scrape.'"[31] Echoing these comments from a 1902 advertisement, Julian Pilling reports being impressed by "how good the tone is. One tends to expect a metallic sound but there is none of this. It is certainly a 'violin like' sound although different from that of a wooden fiddle, a little 'flutey' perhaps."[32] Aleksander Kolkowski, who has recently restored a full Stroviols string quartet, stresses contrast between "the sound of a normal violin" which, "because its under your chin, feels like an extension of yourself," and Stroh sound "coming from the horn," giving "a strange feeling of detachment."[33] He seconds Pilling's assessment of the timbre: "You'd expect them to be tinny, but they are not at all. They sound very flutey, very warm, a little reedy." The sound is, Kolkowski adds, reminiscent of a "very good early gramophone recording." For early-twentieth-century authors promoting the Stroh, the instrument gives traditional string sound a boost, making it richer and bringing it closer to modern ears crowded by a new world of competing entertainment. For the early twentieth-first-century musician engaging with the history of technology there is a contrasting emphasis on the mediation of the violin sound, almost as if the horn represents a hundred-year tunnel for the sound waves to travel through.

These cultural contingencies give the Stroh instruments a complex bearing on the theory of sentics, for the relation of input effort to output sound cannot be straightforwardly assessed. The horn is intended to amplify sound in a particular direction, without a corresponding increase in the performer's effort. This could be identified as a preliminary step toward the sentic age, a categorization that is supported by viewing the Stroh's phonograph technology as a precursor to electrical and ultimately electronic sound recording and reproduction. The "flutey" sound and distancing effect mean that the conventional scale of stringed instrument performance activity yields a sound associated with reduced effort: the "very good early gramophone recording" described by Kolkowski. The gestures associated with playing a gramophone, while considerably greater than those entailed by a CD or MP3, nonetheless constitute a significant reduction from the original effort of playing a stringed instrument. However, when the restored Stroviols quartet plays live, an aesthetic of histrionics begins to emerge: the four musicians are seen to graft as much as—perhaps a shade more than—any classical string ensemble, and certainly a good deal more than is habitual amid the solemnities of avant-garde performance. In the context of new music programming, Kolkowski's restored instruments offer their "reedy" sound as an inversion of the effort paradigm established through music with computers. Close examination of a specific performance helps to demonstrate this point.

Apartment House Stroviols Quartet

Kolkowski's composition *What Hath God Wrought?* plays with the relationship between string and horn features of the Stroviols, including percussive use of the metal, bow grinding, and bowing of horn rims. Not until the third movement are strings bowed, exclusively with harmonics, to create a stylized eeriness. The first and final movements incorporate two Edison cylinder phonographs playing telegraphic sounds, in a tribute to Stroh's work in telegraphy with Charles Wheatstone in the late 1800s. "What hath God wrought?" was the first message transmitted by Samuel Morse in 1844, and in the final movement one cylinder plays this signal while the other reproduces the resonant frequencies produced by bowing each of the four horns (the Morse message and horn frequencies set the rhythmic structure and pitches for the whole composition). In addition, a gramophone plays an old Morse code instructional record from the 1930s. The phonographs and gramophone are connected to the violins and viola with lengths of PVC tubing, so that the recorded sound emerges through their horns (at one point the cello is played through the viola horn using the same technique).

The effort expended here is ultravisible to the audience, and it is coupled with vulnerability of sound output. Messages are subject to blurring, exaggeration, deterioration, crackle, echo. Performance gestures are greater than the output sounds: The plastic tubing at times seems almost alive, with a tendency to escape from where it has been placed, requiring vigilance and repositioning. Themes from *What Hath God Wrought?* were echoed in the other works performed alongside Kolkowski's piece at Kettle's Yard in Cambridge, United Kingdom, in March 2006 (Figure 2.2). Caroline Wilkins's *With Circle and Axis* (originally composed for a traditional string quartet) exploits uncanny effects and ordering, matching the blend of scientific and supernatural interests characteristic of the Stroh period, while *To Be or Not to Only Stand and Wait* by John Lely invokes mechanical toys and the music hall, incorporating kazoos and linking the Strohs to each other once again with tubing. The program concluded with Anton Webern's *Langsamer Satz* (1905), giving the audience a taste of conventional string music played on the Stroviols, with plenty of vibrato and swaying to underline the histrionic aesthetic.

Overall, the Apartment House quartet exploits the Stroviols's ambivalent position in the history of music technology, inviting the audience to encounter phonographic technology on renewed terms. The use of connecting tubes, inspired by early dictation machines and fairground or arcade devices for listening to recordings without the use of horns, additionally functions as a visual parody of the much-less-visible wired and wireless connectivity of electronic devices. This invites self-consciousness about our tendency to measure the past against present levels of technological capacity. The vulnerable dexterity and extra labor on show heighten awareness of instruments as participating in a broader social and material context, creating a thickness to the Stroh sound as it is literally layered with voices from the history of communication technology. While these

FIGURE 2.2
The Apartment House Stroviols quartet performing Aleks Kolkowski's composition *What Hath God Wrought?* at Kettle's Yard in Cambridge, United Kingdom, in March 2006. Courtesy of Aleks Kolkowski.

voices veer toward the Stroviols's scientific associations, a touch of the improvised, irreverent, music-hall style also carries through. A comparable layering effect, similarly tinged with showmanship, is created in Julio d'Escriván's *Ayayay!* through a dialogue between Pianola and iPhone.

Machine Expression: Piano-Players and Player Pianos

Introduced to the market in the late 1890s and declining through the late 1920s, automatic pianos became a symbol of the period's deep ambivalence about human-machine interaction. The basic principle of operation is a perforated roll on which small holes correspond to performance actions (the striking of individual piano notes). In this sense the automatic piano may be compared to an early sequencer.[34] Piano-players are boxes pushed up to the front of a pianoforte, with concealed mechanical fingers striking the keys. Player pianos accommodate the mechanism within the piano body: They may be played automatically by means of a roll, or conventionally by a pianist striking the keys. ("Pianola," a trademark of New York company Aeolian, came to stand loosely for piano-players and player pianos of all makes.) In either case, suction from foot pedaling maintains a vacuum, while the pedal action is also used to wind the roll at a steady rate. Each time a hole passes, air is sucked through to operate a tiny bellows, activating the programmed note. The extent of the hole determines how long the note will last. A bewildering array of variations on this system became available as manufacturers competed for the pinnacle of mechanical ingenuity.[35]

The introduction of a "reproducing piano" by German firm Welte was a turning point in player piano technology. The Welte-Mignon, launched in 1904, entailed the use of "specially perforated rolls that captured the dynamics or 'expression' of a live pianist playing on a special recording piano."[36] The Apollo Marking Piano, invented by Melville Clark in 1912, was a further significant development in piano recording.[37] Before the arrival of electrical recording in the mid-1920s, the reproducing piano was the preferred method of capturing piano performances, with Ignacy Jan Paderewski and Igor Stravinsky among the prominent advocates.[38] As Brian Dolan has documented, however, advertising rhetoric about making the spirit, soul, and expression of accomplished performers accessible to all concealed an industry of human-machine interaction. The recording process required the pianist to play in a more mechanical style than usual, and the resulting rolls had to be treated by an "arranger" to make them sound more "human."[39]

Consumer Music

A further development was the replacement of foot pumping by electrical operation, greatly reducing the skill and effort required to produce music from any roll.[40] Craig Roell identifies this as part of a much broader shift, a "gradual replacement of the Victorian work ethic with the leisure-oriented consumer ethic."[41] Even without electrification, the automatic piano was a symbol of mass entertainment values. Aldous Huxley, in his first novel *Crome Yellow* (1921), used it to heighten his protagonist's suffering at a country house party:

> Denis did not dance, but when ragtime came squirting out of the pianola in gushes of treacle and hot perfume, in jets of Bengal light, then things began to dance inside him. Little black nigger corpuscles jigged and drummed in his arteries. He became a cage of movement, a walking palais de danse. It was very uncomfortable, like the preliminary symptoms of a disease.[42]

As Roell has observed, player pianos were essential to the ragtime "contagion," whereby black music was adopted by white middle-class composers and audiences.[43] By the early 1920s, Huxley was able to treat this established phenomenon with profound cynicism, selecting a foot-pumped Pianola to underline the house party's dependence on a jaded consumer modernity. Denis Stone, an aspiring poet, is taunted in his isolation by the automated culture uniting other guests: "At the pianola, Henry Wimbush, smoking a long cigar through a tunnelled pillar of amber, trod out the shattering dance music with serene patience. Locked together, Gombault and Anne moved with a harmoniousness that made them seem a single creature, two-headed and four-legged."

Fourteen years earlier, Joseph Conrad chose a fully automated model to intrude upon his skulking anarchists in the Silenus beer hall of *The Secret Agent* (1907). Set in 1886, the novel reaches ahead of its period to adopt an instrument that symbolizes its characters' doomed subjection to their own convictions: "An upright semi-grand piano near the door, flanked by two palms in pots, executed suddenly all by itself a valse tune with aggressive virtuosity. The din it raised was deafening."[44] As a vision of pointless destruction is revealed to Comrade Ossipon by the inhumane Professor, the instrument springs once again to life: "The piano [. . .] clanged through a mazurka with brazen impetuosity, as though a vulgar and impudent ghost were showing off."[45] Both authors represent the automatic piano as a source of irritation, a gimmick that removes skill, creativity, and personality from musical performance.

As Dolan and Roell observe, this was a key concern in press coverage, where anxieties about a machine aesthetic were countered by insistence on the human spirit captured in rolls or contributed by the performer.[46] In addition to the control of dynamics afforded by the speed of foot pumping, various devices were incorporated by manufacturers to enhance expression. Operating a player piano with the capacity for dynamic, tempo, and accent control requires considerable practice, and training rolls were available to guide users through staged lessons in their use.[47] "Metrostyle," introduced by Aeolian in 1903, allowed the player to follow a red, wavy line, adjusting the tempo as appropriate.[48] "Themodist" allowed the bass and treble to be accented, giving the music (especially ragtime) an enhanced liveliness. Both innovations could be applied at will by the operator, or automated by the rolls (Themodist uses punched holes to control emphasis; Metrostyle requires attention to the red line and simple movement of a lever). Dolan notes that the "multitude of adaptations and developments [. . .] stimulated much discussion about ways that the machine, hidden within the piano, worked to produce less mechanical-sounding music. Media attention turned to reflect on ways that mechanization might actually embody human dynamic expression or 'feeling.'"[49] Roell also notes the "contradictory ideology" in player piano advertisements, which promoted "ease of play while espousing the individual creativity traditionally associated with the producer ethic."[50] A machine that would at once remove effort and enhance creativity, deploying mechanical genius to become increasingly human: The player piano encapsulates perfectly the combination of faith in technology and mistrust of its applications that characterized the early twentieth century.

The Sweat of a Pianolist

Awareness of effort is central to Rex Lawson's work as a concert pianolist. Born in 1948, Lawson diverted from a traditional career path in music to focus on reproducing pianos and player pianos, performing works by Stravinsky, Percy Grainger, and Conlon

Nancarrow, as well as George Antheil's *Ballet mécanique*.[51] As he will point out in conversation, such works comprise less than 1 percent of the instrument's repertoire, yet these are the pieces called for in concert programs today. On his website and in public talks, Lawson establishes an aesthetics of significant effort for the pianolist, emphasizing that piano-player rolls supply only the notes, leaving dynamics and tempo to be added by the performer. He offers an analogy with the orchestral conductor, who must contribute more than "mere beating of time." A "good conductor" will, he observes, "bring a unified emotional focus to a performance that is being given by a multitude of different personalities. This is usually far better achieved by flexible suggestion than by rigid enforcement, with arm and body movements that dance rather than agitate mechanically."[52]

The contrast between "rigid enforcement" and "flexible suggestion" emerges clearly through different approaches to the forbidding task of realizing *Ballet mécanique*. Antheil's 1924 score called for sixteen player pianos, but as Paul Lehrman explains: "Until the 1990s, this version of the piece had never been performed in its original instrumentation, since the technology for linking and synchronizing multiple player pianos, whether 4 or 16, although theoretically possible when Antheil conceived the piece, turned out not to be practical."[53] In 1999 the piece was performed using MIDI-compatible player pianos, operated by sequencers reading MIDI files.[54] Lawson prefers to work with the 1927 score featuring a solo pianolist, which he performed at Carnegie Hall in July 1989. This version was subsequently performed in Cambridge, United Kingdom, in 2006 and again in 2009, with the Anglia Sinfonia conducted by Paul Jackson. As a reviewer familiar with Lehrman's project noted, "*Ballet mécanique* takes on a whole new degree of light and shade. Yes, it's still a cacophony of 20s *avant-garde* exuberance, but it takes on a good deal of additional subtlety. Lawson feels that the piece is designed to be played on these Edwardian instruments rather than modern digital systems, and that you need to actually *perform* the Pianola part—as he puts it, you need to 'sweat.'"[55]

Perspiration is a recurrent theme in Lawson's comments about Antheil's composition. The Pianola Institute web page strikes a note of regret about the "slightly sanitised flavour" that emerges from the CD recording of the New York 1989 revival. It had been a humid occasion, making for an "atmosphere of healthy sweat [that] suits the *Ballet mécanique* well; machines in the 1920s were not the silent computer driven affairs that we have come to expect, but depended instead on hot oil and coal, just like the steam engines of a past era."[56] Such comments, associating the performer's perspiring body with a steam engine, celebrate the "Victorian work ethic" that Roell sees fading from prominence in the early twentieth century. With an Aeolian piano-player that can be pushed up to any Steinway in any concert hall around the world, Lawson chooses to promote the Pianola as a piece of mechanical ingenuity rather than a precursor to electronic instruments. And he exploits the audience connection afforded by palpable physical effort, citing one particular concert where air was leaking from the Pianola and he had to pedal almost to the point of exhaustion in order to maintain the sound. As sweat poured, the audience applauded with extra gusto. His long beard and lively demeanor prepare the crowd for showmanship, while for the *Ballet mécanique* he is prone to add spats, lending a historical feel. Such strategies make for an accessible and engaging manifestation of the piano-player and its operator as a hybrid human-machine, the positive counterpart of Huxley's "cage of movement" or Conrad's "impudent ghost."

The concept of an automatic piano carries with it the assumption of minimal effort, skill, and dexterity, yielding the equivalent sound output to a pianoforte (if not louder and more intrusive, as the adjectives *shattering* and *deafening* selected by Conrad and Huxley convey). As a concert pianolist, Lawson communicates effort through his physical gestures and skill and dexterity through his website and public demonstrations. Through his showmanship the piano-player is in a continuous process of being curated away from sentics and toward an aesthetic that stops just short of histrionic. Where modernist authors emphasized lack of input effort or skill to signal the hopeless struggle for individual expression against mass culture, Lawson's style rehabilitates the skill and effort involved, offering a chance to revisit the roots of today's consumer culture on renewed terms.

iPhone as Musical Instrument

The 2009 performance of *Ballet mécanique* in Cambridge was the finale in a concert that opened with *Ayayay!*, a new piece for Pianola and iPhone. The composer, Julio d'Escriván, is an exponent of "mixed genre" or "mixed media" (previously known as "soloist and tape") music, in which electronics are combined with traditional instruments. Previous works include *Concerto Demente* (2005), for orchestra and a video game controller soloist. Originally trained as a classical guitarist, he worked for Yamaha in research and development during the 1980s, and he has composed music for film and commercials in America and his native Venezuela. Winner of the Bourges Competition for Electroacoustic Music in 1987 and 1989, he has received numerous prizes for film music in Latin America. He also wrote the music for the film *Balloon* and was awarded a BAFTA for best animation in 1992. Since 2004 he has lectured in creative music technology at Anglia Ruskin University.

The use of iPhone as a musical instrument has been developed by the Stanford Mobile Phone Orchestra (MoPhO, founded in 2007 by Ge Wang, Georg Essl, Henri Penttinen, and Chryssie Nanou) and the Michigan Mobile Phone Ensemble (established in 2009 by Essl).[57] Numerous interfaces for musical and performance application are currently in development.[58] The iPhone's distinctive capability (as compared, for instance, to a Nintendo™ Wii), is its multitouch surface. Both devices incorporate an accelerometer for tracking relative position, but the Wii can only register one button press at a time. The surface of an iPhone or iPad is similar to that of a JazzMutant Lemur, in that multiple touches on the screen can be registered simultaneously. This allows for greater complexity of musical control, making it more like a conventional musical instrument than a games controller: once the iPhone inputs have been programmed to deliver specific sonic effects, the user must devote considerable time and patience to manipulating the multiple parameters and learning how they interact with each other. As d'Escriván explains, the device yields "complex gestures you don't know you can make until you make them."[59]

A popular application for multitouch sound is TouchOSC (Figure 2.3), for sending Open Sound Control messages wirelessly. OSC is a protocol using a "URL-style symbolic naming system" for communication between multimedia devices (for example, computers, sound synthesizers).[60] During performance or rehearsal the device is used in flight-safe mode, so that messages can be sent and received wirelessly without fear of interruption by a phone call. An iPhone running TouchOSC can offer sliders, dials, beat machines, and touchpads to be used singly or simultaneously. D'Escriván works with a "capture and release" paradigm of live electronic music, using the iPhone to capture snatches of Pianola music during the performance, then manipulating them with the touchpad and releasing them during rests in the Pianola part. At the 2009 concert, he paused before playing to explain that one of the iPhone's musical functions in *Ayayay!* was to "subvert what the pianola is doing." A short demonstration of the capture, manipulation, and release process was given. The piece uses granulating software, custom built in SuperCollider in collaboration with Sergio Luque. Granulation is a method for chopping source sound into grains, allowing the sound to be elongated at a chosen moment (for example, the word *cheese* could become "chhhhhhhhhhhhheese" or "cheesssssssssssse").

Playback can also be subject to pitch and dynamic control if the selected device has been programmed appropriately. In the setup used for *Ayayay!*, the sound will become softer if the iPhone is rolled away and louder if it is lifted toward the user. Following the intuitive scheme derived from a piano keyboard, a roll to the left makes the pitch lower and a roll to the right will raise it.[61] While the touchpad controls are sentic, the spatializing and pitch control gestures have an evident correspondence between input effort and output sounds. This correspondence is partly symbolic: the audience will know that the iPhone musician is shaping sound in many more ways than those revealed by the readily apparent set of gestures. It is through what may be described as a qualified sentics that the iPhone enters into dialogue with the Pianola in *Ayayay!* (Figures 2.4 and 2.5).

Ayayay!

Each performance of *Ayayay!* will follow the same basic structure, with some variations in how the iPhone part comes out. The score, which follows a standard concerto model of three movements, indicates fixed notes for the Pianola (these were punched onto three rolls by Lawson). The first movement introduces the iPhone's spatializing gestures, described above. The symbolic importance of spatialization is emphasized by being extended to the orchestra, as waves of sound and movement sweep through the players. Here and in the third movement, the pianolist pedals continuously (rests are included in the roll). The second movement is a conversation between the two soloists, and the

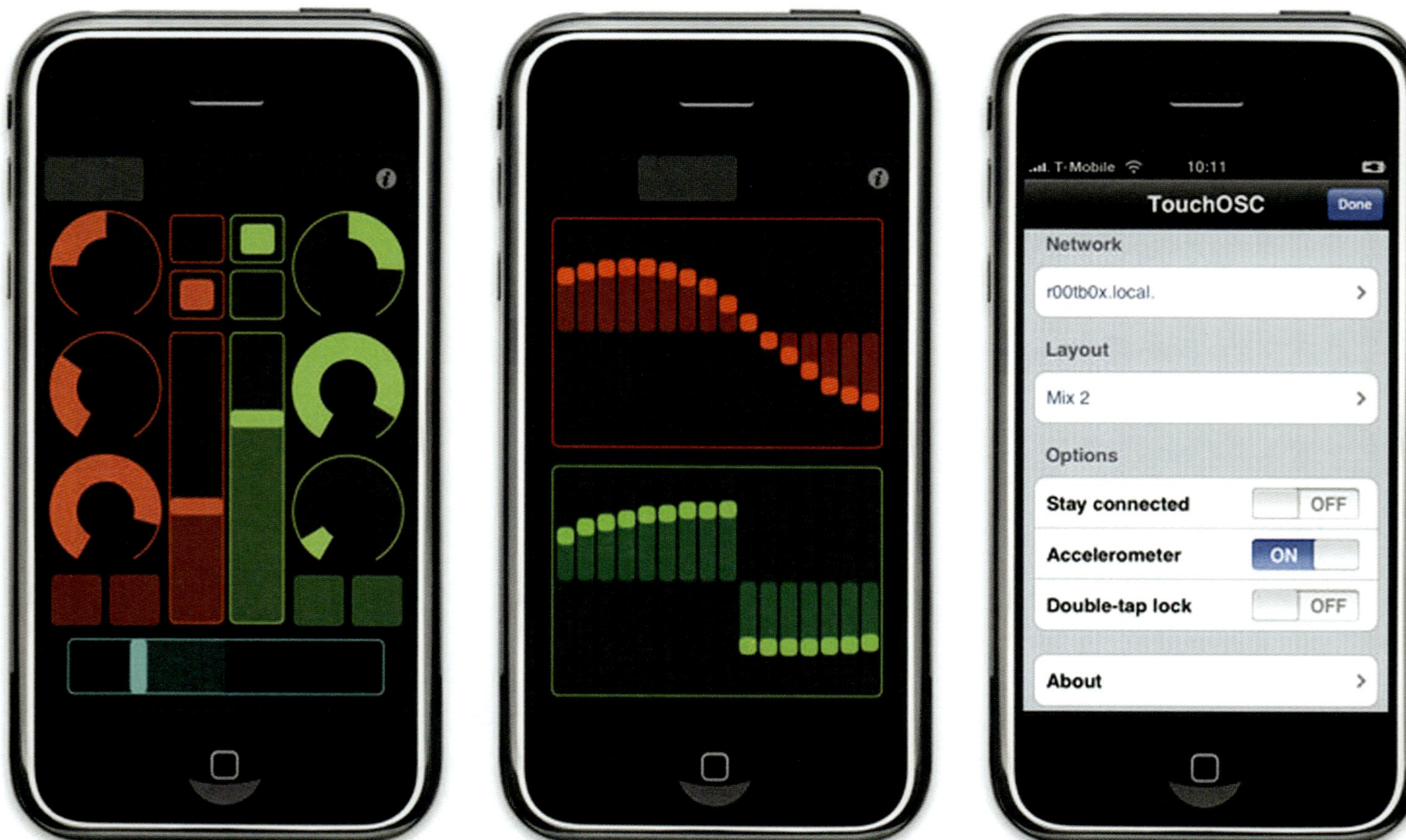

FIGURE 2.3
iPhone screen shots running TouchOSC. Courtesy of Hexler.

pianolist will cease pedaling between portions of his roll to allow the iPhone its response. Throughout the piece, the two soloists are facing each other across the conductor, reinforcing the impression of a dialogue. In the concluding movement the orchestral accompaniment resumes, and the iPhone is more submerged in the music.

In contrast with many mixed-media pieces, the iPhone in *Ayayay!* "has no co-ordinating role and acts as a modifier, in free time, of the material produced by the rest of the ensemble."[62] As Lawson will point out, the Pianola is not an orchestral instrument, and pianolists are generally not used to following conductors. Lawson's expertise makes integration possible, but in performance the combination of Pianola and iPhone creates a tension between the conductor's gestures and overall output sound, stretching the accepted analogy of ensemble as an instrument played by the conductor almost to the breaking point. The two artifacts, from opposite ends of the twentieth century, are not simply bolted on to a conventional ensemble but call into question the very basis on which an orchestra is presumed to operate.

Overall the concerto tells a story about this contemporary device asserting its characteristic gesture set before emerging into the virtuoso foreground and finally being accepted into the mix. *Ayayay!* is a playful title that reflects this allegory of the iPhone becoming an "instrument," punning on the *i* prefix that is becoming ubiquitous throughout Western consumer culture while connoting an exclamation of distress in the composer's native Hispanic discourse. An "ayayay" is also a Mexican love song. Audience members may not consciously realize these connotations, but the title with its exclamation point will nevertheless alert them to a mode of exuberance that is far removed from the stereotype of a laptop artist mostly hidden behind a glowing Apple icon. Like the symbolic use of spatializing gestures, this invites the audience to share in the process of exploring the terms on which a multimedia device can become musically engaging.

The pianolist's aesthetic of significant effort lends a vital historical aspect to the speculation encouraged through *Ayayay!* A triumph of the machine age, the Pianola surprisingly does not represent the obsolescence of effort but rather an amalgam of mechanical and human work in the service of entertainment. When brought into conversation with the iPhone, this lends greater musical complexity and cultural resonance to what would otherwise be a fairly superficial exercise in performing the iPhone's capability as an "instrument." Just as the Pianola was a symbol of the latest technology in its day, the iPhone is one of today's most desirable gadgets. Where modernist authors used the Pianola to vent their antagonism toward mass culture, today's satirical outlets use the iPhone to tease hyperawareness of collusion with consumerism. An *xkcd* character asks whether there is an app for wanting "something more than the pale facsimile of fulfillment brought by a parade of ever-fancier toys? To spend my life restlessly producing instead of sedately consuming?"[63] In 2008 *The Onion* reported "Police: iPhone Left in Hot Car for Three Hours," and in 2009, "Apple Claims New iPhone Only Visible to Most Loyal of Customers."[64] D'Escriván's use of the iPhone as instrument foregrounds the device as a responsive extension of the performer's body, its dialogue with the Pianola

FIGURE 2.4
Rex Lawson on Pianola in Julio d'Escriván's *Ayayay!* at West Road Concert Hall in Cambridge, United Kingdom, in November 2009. Courtesy of Julio d'Escriván.

reminding us that today's concerns about consumer culture have their roots in a preelectronic age.

Conclusion

Both the Apartment House Stroviols quartet and *Ayayay!* demonstrate the potential of artifacts in performance, in particular their scope for creating "bubbles" that can disrupt a linear history of music technology. The implied parallels between mechanical devices and electronic instruments make it harder to identify electronics as "a distinct phase in the development of music." Yet the celebration of mechanical genius as musically interesting in its own right, without being relegated as a precursor to subsequent developments, also makes it harder to categorize electronic instruments as an exaggeration of "existing tendencies." The edging toward histrionics and the compromising of sentics that emerge from these performances encourage audience members to become more self-conscious about their relationship to past and present technology. In particular, the pastiche of familiar performance gestures in the development of musical techniques for the

FIGURE 2.5
Julio d'Escriván with iPhone in his piece *Ayayay!* at West Road Concert Hall in Cambridge, United Kingdom, in November 2009. Courtesy of Julio d'Escriván.

iPhone recalls the human-machine hybridity of devices associated with music technology just prior to electrification. The incorporation of the push-up Pianola in dialogue with the iPhone makes this message explicit, reminding us that present-day identification of musicians as producer-consumers has its roots in an earlier age, with its complex adoration and suspicion of machines. Artifacts on their own also have the potential to trigger responses along these lines, as the Stroviols quartet demonstrates through the use of supporting contextual materials (recordings and tubing) that create an implicit dialogue between preelectronic instruments and electronic devices.

Such encounters are made possible through the commitment of collector-performers and collector-composers, who give their chosen artifact immediacy for today's audiences. Their performances are carefully worked out with both audience and history in mind, renegotiating the experience of authenticity in audience encounters with past and current objects.[65] Committed to restoration and collaboration, these live curators lend visual and aural form to compelling questions of cultural production that might otherwise remain corralled on the pages of theoretical texts, and by doing so they widen the circle and terms of debate about technology and music in our lives.

Notes

I thank Aleks Kolkowski and Julio d'Escriván for providing extensive information about their instruments and compositions, Paul Jackson for commenting on a draft, and Duncan Miller for giving me a lesson on his player piano.

1. Hans-Joachim Braun, "Introduction," in *Music and Technology in the Twentieth Century*, ed. Hans-Joachim Braun (Baltimore, MD: The Johns Hopkins University Press, 2002), 9.

2. Alison Rabinovici, "Augustus Stroh's Phonographic Violin. A Journey: Victorian London, Australia, Transylvania," *Galpin Society Journal* 58 (2005): 100–23, 101; Brian Dolan, *Inventing Entertainment: The Player Piano and the Origins of an American Musical Industry* (Lanham, MD: Rowman & Littlefield, 2009), 156–57.

3. Julio d'Escriván, "To Sing the Body Electric: Instruments and Effort in the Performance of Electronic Music," *Contemporary Music Review* 25.1/2 (2006): 183–91, 189, 185.

4. D'Escriván, "To Sing the Body Electric," 183.

5. D'Escriván, "To Sing the Body Electric," 184.

6. D'Escriván, "To Sing the Body Electric," 184, 188.

7. D'Escriván, "To Sing the Body Electric," 190.

8. David Toop, *Haunted Weather: Music, Silence and Memory* (London: Serpent's Tail, 2004), 14.

9. Paul Théberge, *Any Sound You Can Imagine: Making Music/Consuming Technology* (Hanover, NH: Wesleyan University Press, 1997), 199.

10. Théberge, *Any Sound You Can Imagine*, 35, 213.

11. Théberge, *Any Sound You Can Imagine*, 83.

12. Théberge, *Any Sound You Can Imagine*, 230, 241, 244.

13. Murray Schafer, *The Tuning of the World* (New York: Knopf, 1977), 90.

14. Roland Barthes, *Image, Music, Text*, trans. Stephen Heath (London: Fontana, 1977), 182.

15. Barthes, *Image, Music, Text*, 183.

16. Barthes, *Image, Music, Text*, 182.

17. D'Escriván, "To Sing the Body Electric," 185–87.

18. T. Pinch, and F. Trocco, "The Social Construction of the Early Electronic Music Synthesizer," in *Music and Technology in the Twentieth Century*, pp. 67–83, ed. H-J. Braun (Baltimore, MD: The Johns Hopkins University Press, 2002), 72.

19. Théberge, *Any Sound You Can Imagine*, 173.

20. On the history of Stroh instruments, see Rabinovici. An informative exhibition on "Horn Amplified Strings" curated by Aleks Kolkowski was displayed at the Musical Museum, Brentford, UK, October 2009.

21. Rabinovici, "Augustus Stroh's Phonographic Violin," 100.

22. Rabinovici, "Augustus Stroh's Phonographic Violin," 101–2.

23. Rabinovici, "Augustus Stroh's Phonographic Violin," 102–4.

24. Kolkowski, "Horn Amplified Strings."

25. Rabinovici, "Augustus Stroh's Phonographic Violin," 104.

26. Kolkowski, "Horn Amplified Strings."

27. Rabinovici, "Augustus Stroh's Phonographic Violin," 114.

28. Rabinovici, "Augustus Stroh's Phonographic Violin," 112.

29. "The Stroh Violin," *Bazaar, The Exchange and Mart*, November 27, 1901, quoted in Rabinovici, "Augustus Stroh's Phonographic Violin," 107.

30. Craig Roell, *The Piano in America, 1890–1940* (Chapel Hill: University of North Carolina Press, 1989), 155.

31. *Strand Magazine*, 1902, quoted in Julian Pilling, "Fiddles with Horns," *Galpin Society Journal* 27 (1975): 86–92, 88.

32. Pilling, "Fiddles with Horns," 89.

33. Quoted in Charlotte Higgins, "Brass and Strings' Secret Love Child," *The Guardian*, November 2, 2006. http://www.guardian.co.uk/music/2006/nov/03/classicalmusicandopera (accessed April 30, 2010).

34. Théberge, *Any Sound You Can Imagine*, 223; Paul Jackson, "'. . . Out of and for Machines . . .': Time-Space and Time-Form in George Antheil's *Ballet mécanique*," in *Sonic Ideas/ Ideas Sónicas* 2(2), Centro Mexicano para la Música y las Artes Sonoras, 14–15.

35. On the history and technology of automatic pianos see Dolan, *Inventing Entertainment* and Roell, *The Piano in America*.

36. Dolan, *Inventing Entertainment*, 46.

37. Dolan, *Inventing Entertainment*, 51.

38. Dolan, *Inventing Entertainment*, 92–93; Roell, *The Piano in America*, 44, 48.

39. Dolan, *Inventing Entertainment*, 84–85, 129–30, 146–50, 159–60.

40. Dolan, *Inventing Entertainment*, 154.

41. Roell, *The Piano in America*, 155.

42. Aldous Huxley, *Crome Yellow* (London: Granada, 1982), 50.

43. Roell, *The Piano in America*, 32–36; see also Dolan, *Inventing Entertainment*, 101.

44. Joseph Conrad, *The Secret Agent* (Mineola, NY: Dover, 2001), 38. On the player piano modernist literature see Tim Armstrong, "Player Piano: Poetry and Sonic Modernism," *Modernism/ Modernity* 14.1 (2007): 1–19.

45. Conrad, *The Secret Agent*, 42.

46. Roell, *The Piano in America*, 53–59; Dolan, *Inventing Entertainment*, chapter 8.

47. Dolan, 116. I am grateful to Duncan Miller for introducing me to the training roll on his player piano.

48. Dolan, *Inventing Entertainment*, 48.

49. Dolan, *Inventing Entertainment*, 51.

50. Roell, *The Piano in America*, 157.

51. R. Lawson, "Biography," *Rex Lawson–Concert Pianolist*. http://www.rexlawson.com (accessed April 30, 2010).

52. R. Lawson, "What Is a Pianolist?" *Rex Lawson–Concert Pianolist*. http://www.rexlawson .com (accessed April 30, 2010).

53. P. D. Lehrman, "Ballet mécanique," *The Ballet mécanique page*. http://www.antheil.org/, and R. Elen, "Ballet mécanique in Cambridge," *Brideswell Associates*. http://brideswell.com/content/ technology/ballet-mecanique-in-cambridge/ (accessed January 9, 2010).

54. P. D. Lehrman, "The Schirmer Project," *The Ballet mécanique page*. http://www.antheil .org/ (accessed January 9, 2010).

55. Elen, "Ballet mécanique in Cambridge."

56. "Pianola Institute Concerts–Archive," *The Pianola Institute*. http://www.pianola.org/concerts/ concerts_archive.cfm (accessed April 30, 2010).

57. *Stanford Mobile Phone Orchestra (MoPhO)*, http://mopho.stanford.edu/; *The Michigan Mobile Phone Ensemble*, http://mopho.eecs.umich.edu/ (accessed April 30, 2010).

58. Peter Kirn, "iPad, iPhone + Music Weekend Update: Logic, Live Control, or All-in-one Music," *Create Digital Music*. http://createdigitalmusic.com/2010/04/11/ipad-music-weekend-update -ableton-live-control-or-all-in-one-music/ (accessed April 30, 2010).

59. In conversation April 22, 2010.

60. "Introduction to OSC," opensoundcontrol.org. http://opensoundcontrol.org/introduction-osc (accessed April 30, 2010).

61. See Tellef Kvifte's contribution to the present volume.

62. P. Jackson, personal communication, May 5, 2010.

63. "iPhone or Droid," *XKCD: A Webcomic of Romance, Sarcasm, Math, and Language*. http://xkcd.com/662/ (accessed April 30, 2010).

64. "Police: iPhone Left in Hot Car for Three Hours" and "Apple Claims New iPhone Only Visible to Most Loyal of Customers," *The Onion*. http://www.theonion.com/articles/police-iphone-left-in-hot-car-for-three-hours,2575/andhttp://www.theonion.com/articles/apple-claims-new-iphone-only-visible-to-most-loyal,2772/ (accessed April 30, 2010).

65. For a discussion of authenticity in relation to a sound installation that does similar work with war memory instead of artifacts, see R. Benschop, "All the Names: Soundscapes, Recording Technology, and the Historical Sensation," in *Sound Souvenirs: Audio Technologies, Memory and Cultural Practices*, pp. 182–98, ed. K. Bijsterveld and J. van Dijck (Amsterdam: Amsterdam University Press, 2009).

Bibliography

Armstrong, Tim. "Player Piano: Poetry and Sonic Modernism." *Modernism/Modernity* 14.1 (2007): 1–19.

Barthes, Roland. *Image, Music, Text.* Translated by Stephen Heath. London: Fontana, 1977.

Benschop, R. "All the Names: Soundscapes, Recording Technology, and the Historical Sensation." In *Sound Souvenirs: Audio Technologies, Memory and Cultural Practices*, pp. 182–98, ed. K. Bijsterveld and J. van Dijck. Amsterdam: Amsterdam University Press, 2009.

Braun, Hans-Joachim. "Introduction." In *Music and Technology in the Twentieth Century*, pp. 9–32, ed. Hans-Joachim Braun. Baltimore, MD: The Johns Hopkins University Press, 2002.

Conrad, Joseph. *The Secret Agent.* Mineola, NY: Dover, 2001.

d'Escriván, Julio. "To Sing the Body Electric: Instruments and Effort in the Performance of Electronic Music." *Contemporary Music Review* 25.1/2 (2006): 183–91.

Dolan, Brian. *Inventing Entertainment: The Player Piano and the Origins of an American Musical Industry*. Lanham, MD: Rowman & Littlefield, 2009.

Elen, R. "Ballet mécanique in Cambridge." *Brideswell Associates*. http://brideswell.com/content/technology/ballet-mecanique-in-cambridge/ (accessed January 9, 2010).

Higgins, Charlotte. "Brass and Strings' Secret Love Child." *The Guardian*, November 3, 2006. http://www.guardian.co.uk/music/2006/nov/03/classicalmusicandopera (accessed January 9, 2010).

Huxley, Aldous. *Crome Yellow*. London: Granada, 1982.

Jackson, Paul. "'. . . Out of and for Machines . . .': Time-Space and Time-Form in George Antheil's *Ballet mécanique*." In *Sonic Ideas/Ideas Sónicas* 2(2). Centro Mexicano para la Música y las Artes Sonoras.

Kolkowski, Aleks. "Horn Amplified Strings." Musical Museum, Brentford, October 2009.

Lawson, R. "Biography" and "What Is a Pianolist?" *Rex Lawson–Concert Pianolist*. http://www.rexlawson.com (accessed April 30, 2010).

Lehrman, P. D. "Ballet Mécanique." *The Ballet mécanique page*. http://www.antheil.org/ (accessed January 9, 2010).

Pilling, Julian. "Fiddles with Horns." *Galpin Society Journal* 27 (1975): 86–92.

Pinch, T., and Trocco, F. "The Social Construction of the Early Electronic Music Synthesizer." In *Music and Technology in the Twentieth Century*, pp. 67–83, ed. H-J. Braun. Baltimore, MD: The Johns Hopkins University Press, 2002.

Rabinovici, Alison. "Augustus Stroh's Phonographic Violin. A Journey: Victorian London, Australia, Transylvania." *Galpin Society Journal* 58 (2005): 100–23.

Roell, Craig. *The Piano in America, 1890–1940*. Chapel Hill: University of North Carolina Press, 1989.

Schafer, Murray. *The Tuning of the World*. New York: Knopf, 1977.

Théberge, Paul. *Any Sound You Can Imagine: Making Music/Consuming Technology*. Hanover, NH: Wesleyan University Press, 1997.

Toop, David. *Haunted Weather: Music, Silence and Memory*. London: Serpent's Tail, 2004.

CHAPTER 3

Technology and Authenticity

The Reception of the Hammond Organ in Norway

Frode Weium

Introduction

The advent of new music technologies has always sparked debate. Some people have been enthusiastic about the possibilities afforded by technology to create new sounds. The claim has also been made that technology has promoted democratization by making it easier for people to create their own music. Others, however, have been concerned that technology could become more important than human creativity. Technology has often been viewed as something artificial and false, as something that is not a true part of the musical domain. New musical devices have been rejected as mere machines, differing from conventional musical instruments, which are considered to be more real and authentic. Such debates have often concerned where to draw the boundary between instruments and machines. As pointed out by Trevor Pinch and Frank Trocco, this boundary has been repeatedly redrawn.[1]

"Never in musical history has any new instrument so captured the attention and curiosity of the public,"[2] one of the early advertising leaflets from the Hammond Clock Company in Chicago stated. Even though the claim is disputable, the attention that the Hammond organ received during the years following its introduction in 1935 illustrates different reactions to new music technology. Some celebrated the new organ as a technical sensation and a revolutionary instrument. It was also one of just a few electronic—or more strictly speaking, electric or electromechanical—musical instruments developed before the Second World War that became a commercial success. But there were others who condemned the Hammond organ as a false imitation of the pipe organ, and they opposed the use of the term *organ* in its name and its description.

The first part of this chapter gives an account of the early history of the Hammond organ, looking at why it became a success and focusing on early conceptions of the newcomer. Of particular interest are the so-called Hammond case in the United States (1936–1938) and the position taken by the Catholic Church toward this new instrument. The larger part of the chapter examines the reception of the Hammond organ in Norway from the mid-1930s up to the 1960s. It describes how the meaning of the instrument was established by different social groups, and it argues that it was defined primarily as a church instrument. Special attention is given to some particular organs imported by Norway during this period and how these instruments were adapted to different environments. Furthermore, the chapter shows that the first enthusiasm in the 1930s turned into hostility toward the Hammond organ, and it focuses on the notion of the instrument as an artificial surrogate for the pipe organ rather than as an authentic musical instrument in its own right. Finally, the reception of the Hammond organ is related to more general conceptions of technology and authenticity.

Much of this chapter is based on attempts made by me to identify, trace, and study the Hammond organs imported by Norway during the first decades. This work has been done with the aid of archival sources at the Norwegian Museum of Science and Technology and several informants throughout the country.[3] In general, the Hammond organ has received little attention from either the field of the history of technology or the field of music history. Most standard histories of electronic music only mention this organ in brief passages, as part of the prehistory of the developments after the war, or when referring to popular music. What has been written about the instrument has mainly been undertaken by Hammond enthusiasts out of their love for the organ. These writings tend to focus on legendary models such as the B-3 introduced in 1955, and the use of the organ in jazz, blues, pop, and rock music by well-known musicians such as Jimmy Smith, Keith Emerson, and John Lord.

Conservative Ambitions

Between 1920 and 1950 well over a hundred electric, electromechanical, electroacoustic, and electronic musical instruments were invented. They had unfamiliar names such as Hellertion, Elektrochord, and Ondioline, to mention but a few examples. Only a handful of these instruments achieved any commercial success or acceptance among professional musicians. The Hammond organ belongs to the exceptions. It not only became commercially successful in its own time; the Hammond tonewheel organ is still popular even today.[4]

Laurens Hammond (1895–1973) was a nonmusical maker of clocks from Chicago with a background in engineering and science. He patented his electric organ in 1934, and the first Model A was introduced the following year at the Industrial Arts Exposition in Rockefeller Center in New York (Figure 3.1). Within a few months, more than 1,400

FIGURE 3.1
Laurens Hammond and his Model A organ, 1935–1936. Courtesy of the Chicago History Museum (ICHi-51842). Photographer: Kaufmann & Fabry.

orders were received, and during the first three years of business around five thousand instruments were sold.[5] Initially the Hammond organ was designed for church use and marketed as the heir to the pipe organ. However, a significant number were soon also sold to private homes. Among the well-known early purchasers were Canterbury Cathedral, the Boston Symphony Orchestra, George Gershwin, and Henry Ford. The founder of the Ford motor company showed a great interest in the development of the Hammond organ and placed his first order before production started.

Hammond has been characterized as "the Henry Ford of the music business."[6] There were several reasons why his organ became a commercial success. Compared with the pipe organ it was easier to install, more portable, and less demanding of space. Just as important was the low cost. The Model A started at $1,250, including a tone cabinet, while a pipe organ could often cost ten times as much. Furthermore, the expenses of upkeep were considerably lower for the Hammond. As argued by an enthusiastic American organist: "There is no installation expense, the maintenance cost is trifling, and the instrument is always in tune and ready to meet all demands made upon it."[7] The Hammond organ could also compete with the grand piano. Advertisements stated that even the average home could afford such an organ: "Fine organ music is no longer the exclusive privilege of wealth."[8]

The Hammond Clock Company, which changed its name to the Hammond Instrument Company in 1937 and the Hammond Organ Company in 1953, presented the organ as a technical revolution and as the result of scientific work and the mind of a genius. A leaflet from the U.K. agent Boosey & Hawkes compared the organ to other important inventions:

> New chapters in our social history were opened when, two or three generations ago, the telephone and the typewriter arrived on the scene. There can be no doubt whatever that the advent of this new instrument must have an equally startling significance in the world of music. Science and the laws of dynamics have been . . . harnessed to the inventor's will.[9]

However, the simplicity of the Hammond organ is striking. As Hans-Joachim Braun has pointed out: "Compared with other electronic organs of the time Hammond's instrument had the advantage of mechanical simplicity which made it suitable for mass production."[10] The key innovations were the tonewheel and the drawbar. The instrument had ninety-one small tonewheels turning at a constant speed by means of a synchronous motor and revolving close to coil-wound permanent magnets. Each wheel was notched with differently shaped teeth, which created variations in the magnetic field, producing an electrical impulse and a fundamental musical tone. By pulling drawbars mounted above the two manuals, harmonics were added to form the color of the tone (so-called additive timbre synthesis). One could "mix beautiful tone colors as an artist mixes the paints on his palette," an instruction manual for harmonic drawbars announced.[11]

In order to demonstrate to customers how the Hammond organ worked, the company made small models showing the tonewheel and the drawbar system. The model in Figure 3.2 ran on a battery and could be attached to a speaker. Two tonewheels with differently shaped teeth and two electromagnetic pickups are seen on the left, while two harmonic drawbars that can be pulled out to eight notched positions are mounted on the top.

The tonewheels used by Hammond were similar to the ones used by the American inventor Thaddeus Cahill in his Telharmonium at the end of the nineteenth century. The Telharmonium was an enormous instrument weighing two hundred tons and using spinning tone generators to create musical sounds transmitted over wires and heard through telephone receivers. However, Hammond had the benefit of vacuum-tube amplification, so his tonewheel generators could be much smaller. The motor, which was synchronized to the frequency of the electrical supply and stabilized the tone, had been developed by Hammond himself in the 1920s to power his electric clocks. During the Great Depression he had started looking for other products that could use his synchronous motor. This may be one of the reasons why he "turned his non-musical mind to the invention of electric music."[12] Technologically speaking, the Hammond organ was a characteristic instrument for a period that combined mechanical, electromagnetic, and electronic technologies. The tonewheel organ even outlived this period, remaining basically the same instrument until the last models were produced in 1974.[13]

Another interesting point to be made is that Hammond's ambitions might—after all—be said to have been relatively conservative. According to Paul Théberge, "He did not set out to create a revolutionary new instrument but simply to design a more modern and cost-efficient organ."[14] In broader terms Barry Schrader has commented that, "Historically, the most successful electro-acoustic instruments are those that most closely resemble conventional acoustic instruments in both performance design and sound production."[15] In appearance the Hammond organ was a familiar keyboard instrument. The step away from earlier keyboard instruments was not too extensive and did not require major reorientation for the performer.

This point becomes apparent if one compares the Hammond organ with other electronic instruments from the period, such as the Theremin, the Ondes Martenot, or the Trautonium. The example most illustrative is probably the Theremin, a radiolike electronic instrument developed around 1920 by the Russian inventor Leon Theremin. In 1929 he signed an agreement with the Radio Corporation of America (RCA) for commercial production of the instrument. It was presented as a perfect home instrument that anyone could easily learn to play. However, fewer than five hundred instruments were sold. The most obvious reasons for this failure were the unconventional design and the unfamiliar playing technique required. The Theremin was played by moving the hands around two antennas, without any direct physical contact between the player and the instrument. Thus, players lacked a point of reference, something that could help them locate single notes.[16]

FIGURE 3.2
Model (ca. 1960) showing the tonewheel and the harmonic drawbar system. Courtesy of the Norwegian Museum of Science and Technology.

Théberge has also noted that "Hammond was as skillful in marketing and promoting his instruments as he was in designing them."[17] Extensive advertising appeared in home consumer magazines and organ trade publications. Paid announcements updated the readers on how many organs had been installed in churches and concert halls, and they referred to statements made by famous musicians and composers. Even magazines that otherwise took a critical stance toward the Hammond organ—such as the *Diapason* and the *Caecilia* in the United States—published the advertisements. After the Second World War Hammond's marketing strategies enjoyed great success in the lucrative home market. Popular home models including the small spinet organs, as well as a series of innovations such as vibrato, percussion effects, and chorus were introduced. The instrument became a household word. People often just said "Hammond" when they referred to an electronic organ.

The Hammond Case and the Catholic Ban

The Hammond organ was not the only electric or electronic organ developed in the United States in the 1930s. Richard H. Ranger's Rangertone and Ivan Eremeeff's electronic organ were introduced at about the same time. However, the Hammond organ represented the first real threat to the traditional pipe organ industry, which according to Orpha Ochse's historical examination of the organ in the United States was already "struggling for existence during the depression."[18] American organ builders and organists soon expressed concerns about the Hammond organ and were provoked by the company's advertisement campaigns.

In 1935 the American Guild of Organists filed complaints with the Federal Trade Commission, questioning the use of the word *organ* and charging Hammond with unfair sales practices and methods of competition. A case was opened in 1936 when the Pipe Organ Manufacturers Association made a new complaint. Major objections were concerned with advertisements that claimed that "real organ music of unbelievable beautiful quality is now possible in any home at an expense no greater than that of a good piano," that the Hammond organ "produces the entire range of tone coloring necessary for the rendition, without sacrifice, of the great works of classical organ literature," and that many organists agreed it was comparable to a $10,000 pipe organ. In short, the organ builders and the organists denied that the Hammond organ could serve a church as well as the genuine article.[19]

Several hearings and examinations included a remarkable comparison test in the chapel of the University of Chicago, where a panel of professional musicians and university students were invited to tell the difference between a $2,640 Hammond Model A organ and a $75,000 Aeolian-Skinner pipe organ. The Hammond tone cabinets were concealed amid the pipes, and the consoles of both organs were hidden behind screens. A number of compositions were played by the organist and Hammond employee, Porter W. Heaps. While the students were wrong half the time, the experts did better, but neither of the groups was without error.

The debates were still heated as the Hammond case moved toward an end in 1938. Hammond's attorney predicted the demise of the pipe organ industry, just as the making of kerosene lamps had already vanished in the age of electricity. Finally, the Federal Trade Commission decided that the Hammond Instrument Company had to stop publishing advertisements claiming that the organ could produce an infinite number of tone variations. They had to restrict their claims to a total of 253 million possible tones! However, the company was permitted the use of the term *organ*, and by the end of the trial Hammond had been given valuable publicity that probably contributed to extra sales.[20]

The Hammond organ also met opposition from other institutions. In December 1938 and in September 1939 the Holy See in Rome decreed "that the Hammond should not be introduced into churches."[21] This might have been partly a response to the establishment of an Italian distributor of Hammond organs, the company Microtecnica in Turin. After the war electronic organs were permitted to replace pipe organs that had been destroyed in the bombing of European cities, but officially the Catholic ban of such organs continued.

In the early 1960s the Second Vatican Council and the Constitution on Sacred Liturgy reaffirmed that pipe organs were preferable: "In the Latin Church the pipe organ is to be held in high esteem, for it is the traditional musical instrument which adds a wonderful splendor to the Church's ceremonies and powerfully lifts up man's mind to God and to higher things." However, electronic organs could be admitted "on condition that the instruments are suitable, or can be made suitable, for sacred use."[22] Shortly afterward, Pope Paul VI donated twenty-two electronic organs to Italian churches. Electronic organs from Hammond, Baldwin, Wurlitzer, and other manufacturers were also installed throughout other European countries. In Germany, for instance, the number of electronic organs in Catholic churches and chapels had by 1972 reached 1,500.[23]

In general, the resentment toward the Hammond organ lasted longer in Europe than in the United States. A battle similar to the one between Hammond and the American organ builders in the 1930s took place in Germany in the 1960s between the electronic organ manufacturer Ahlborn Orgel and the Bund Deutscher Orgelbaumeister. The latter was supported by the Gesellschaft der Orgelfreunde, a society consisting of Evangelical and Catholic Church musicians as well as pipe organ manufacturers. They regarded the electronic organ as a false imitation of the pipe organ, and they fought against the use of the word *organ* in the naming of such instruments. In 1964 they published a collection of essays in which they instead proposed the term *elektrium*. However, they turned out to be no more successful than the American organ builders had been some decades earlier.[24]

The Hammond Organ Arrives in Norway

The small company of Arvid Dahm in Oslo, Norway, was among the first to sell Hammond organs outside the United States. Arvid Dahm had started his business in 1932, selling synchronous electric clocks for the Hammond Clock Company (a smaller clock model from the collection of the Norwegian Museum of Science and Technology can be seen in Figure 3.3). Dahm's education as an engineer and his earlier experience of installing telephone centrals for the Automatic Telephone Company in Hammond's hometown of Chicago helped convince the company that their organ sales would be in safe hands.

The first Hammond organs arrived in Oslo in December 1936. During the early years Dahm also distributed organs in Sweden and Denmark. Compared with the United States, however, sales in Scandinavia were very modest. No more than twenty Hammond organs were sold in Norway before the war, and during the German occupation imports stopped completely. After the war, sales slowly picked up, and by the early 1950s the number of Hammond organs in Norway had doubled.[25] Even though few organs were imported, the instrument received much attention. Heated discussions reflected different views of the Hammond organ and of new music technology in general.

In February 1937 the Hammond organ was demonstrated to the Norwegian press. The prominent organist and cantor Arild Sandvold played music from Wagner's

FIGURE 3.3
Synchronous electric clock from the 1930s, produced by the Hammond Clock Company and sold by Arvid Dahm. Courtesy of the Norwegian Museum of Science and Technology.

Tannhäuser. According to Sandvold, who was very enthusiastic, the organ was "epoch-making."[26] The press was also overwhelmed by this "miraculous Hammond organ from America,"[27] as one newspaper called it. The emphasis placed on the origin of the instrument was hardly coincidental, the United States being associated with innovation and progress. Another newspaper called the Hammond organ "one of the technical wonders of our time" and assured readers that the instrument in no way made the artist behind the keyboard superfluous.[28] The newspapers could also tell that several Hammond organs were now on their way to Norwegian churches, assembly buildings, and private homes.

In November the same year a Hammond organ became the main attraction at a concert given by the Oslo Philharmonic Society Orchestra. The concert was held at the University of Oslo and broadcast to a wider audience. Sandvold was organ soloist, while the orchestra was led by the internationally known conductor Odd Grüner-Hegge. The program included works by Händel and Boëllmann. This time, however, reactions differed. One journalist pointed to the great historical importance of the instrument: "There seems to be no doubt that it deserves the sensational attention it has aroused. It turns all inherited conceptions about the concept of the organ upside down."[29] Others were far more critical: "Does this instrument have any pure tones at all?" another journalist asked.[30] When the periodical of the Norwegian Association of Organists reprinted a text from the concert program that described the Hammond organ as every organist's dream, one of its readers found this totally unacceptable and asked how any organist could accept such an unworthy instrument.[31]

The Oslo Philharmonic used a Hammond organ for many years. Grüner-Hegge also conducted the orchestra with the organ for the return of King Haakon to Norway after the liberation in 1945. Some years later he appeared in the *Hammond Times*, a newsletter published by the Hammond Organ Company, with a large, front-page picture in which he was portrayed with the organ. "I would like to give the Hammond organization my best compliments for creating such an all-through excellent instrument," he announced.[32]

Unlimited Possibilities

Several relevant social groups participated in establishing the meaning of the Hammond organ in Norway.[33] Engineers were among those who showed the strongest interest in electronic musical instruments during the 1920s and 1930s. The House of Engineers hosted demonstrations of instruments such as the Theremin, the Trautonium, and the Neo-Bechstein piano, and the engineering magazines *Teknisk Ukeblad* and *Tidens Teknikk* described these instruments through enthusiastic articles. In May 1937 *Teknisk Ukeblad* presented the Hammond organ to its readers in these terms: "[Now there is] a new and epoch-making invention in the field of electronic music." The magazine emphasized that the possibilities of the instrument were "so to speak unlimited."[34] In the same year the engineer Georg Brochmann presented the Hammond organ, together with other electronic instruments, in his best-selling book *Mennesket og maskinen* (*Man and the Machine*). "It is obvious," he wrote, "that instruments with such fantastic opportunities open up completely new perspectives for the development of music."[35] Brochmann also noted that the Hammond organ could produce the tone colors of very different musical instruments. In other words, Brochmann and the Norwegian engineers perceived the Hammond organ as not just an imitation of the traditional pipe organ, but rather as a completely new and modern instrument with rich possibilities.

The Norwegian Broadcasting Corporation was often the first to introduce new music and sound technologies, and it played an important role in making the Hammond organ known to a wide audience. A Hammond Model BC with a C-40 tone cabinet was ordered in September 1937, and a Model E with a D-20 tone cabinet in July 1938. The organs were used by the Radio Orchestra, in devotions and as solo instruments. The Radio Orchestra was quite small. Many instruments were lacking, and these were often substituted with imitated sounds from the Hammond organs. When, for instance, Ravel's *Bolero* was to be performed, the organs were used to play the difficult trombone solo.[36]

After the war, when the Norwegian Broadcasting Corporation moved into new buildings, both organs were refurbished in a completely new design. The Hammond E model was initially designed to harmonize with church interiors and was constructed to meet the demands of church organists. It was built with a walnut case in Gothic style, and the preset keys had a round and more traditional "typewriter" design than they had in other models. As put by one commentator, "It is the Hammond E that represents the perfect symbiosis of religious tradition and secular innovation!"[37] In order to make the Hammond E organ from 1938 fit in better with the modern interiors of the studios of the Norwegian Broadcasting Corporation, a new wooden case was made out of oak instead of walnut. It had plain side panels, a sunblind-like back panel that could be pulled down to cover the manuals, and an integrated chair. The work was done by one of Dahm's engineers, and a completely new tone cabinet with a similar design was also made. Among others, Odd Grüner-Hegge used the organ in the 1950s (Figure 3.4). However, at some point, the organ was given to a private owner, and the distinctive wooden case was replaced by a third and more traditional-looking walnut case from a Hammond Model H-112 organ, which was probably better suited to a private home (Figure 3.5). Only the tone cabinet kept the same unique design. Today the 1938 Model E organ is part of the collection of the Norwegian Museum of Science and Technology. It may serve as an example of how the Hammond organ could be adapted to different environments.[38]

In Norway it was not really until the 1960s that the Hammond organ became popular as a home instrument. Very few of the early organs were bought by amateurs. Neither were there many such instruments in restaurants or similar places before the 1960s. In the United States the Hammond organ was used by jazz bands and dance orchestras, in night clubs and theaters, as early as the 1930s. In Norway this was hardly the case.

FIGURE 3.4
Odd Grüner-Hegge at the rebuilt Hammond Model E organ of the Norwegian Broadcasting Corporation (ca. 1951). Photo: NTB Scanpix.

FIGURE 3.5
The 1938 Hammond Model E organ and the tone cabinet as they are today. Courtesy of the Norwegian Museum of Science and Technology.

The exception is a Hammond Model BC Lafleur organ used in the popular dance restaurant Regnbuen in Oslo. The Lafleur models were built under license by J. R. Lafleur & Son (owned by Boosey & Hawkes) in London between 1936 and 1939. The physical appearance of these organs was well suited to cinemas, theaters, and entertainment venues. The restaurant Regnbuen acquired its Lafleur organ when it opened in 1938. The art deco design and the large lily symbol in front gave the organ a distinctive look (Figure 3.6), quite different from the more common church models. When the organ was replaced in the mid-1960s it was, nevertheless, moved to a church. Later on it was sold to a private collector, and today it is on display at the Norwegian Museum of Science and Technology.[39]

Norwegian jazz musicians initially showed little interest in the Hammond organ. The first documented uses within jazz dates from 1957 and 1959.[40] During the 1960s the instrument was taken up by some jazz and rhythm & blues bands, and toward the end of the decade legendary American jazz organists such as Jimmy McGriff and Jimmy Smith visited Norway. By that time several new Hammond organ models with a design less adapted to churches had come on the market, as well as electronic organs from Baldwin, Lowrey, and Wurlitzer.

FIGURE 3.6
Hammond BC Lafleur organ from 1938 used in the restaurant Regnbuen, Oslo. Courtesy of the Norwegian Museum of Science and Technology.

A Church Instrument

Despite the use of the Hammond organ by the Oslo Philharmonic, the Norwegian Broadcasting Corporation, the restaurant Regnbuen, and eventually some jazz musicians,[41] in Norway it was essentially a church instrument, rather than being associated with concerts, dance halls, or jazz clubs. Organists, priests, and congregations were the most important groups defining the Hammond organ. In 1939 Arvid Dahm concluded that "the completely electric organ has now attained general and undisputable status as a church instrument."[42] He could present recommendations from several organists and priests in Norway and Sweden, who described the organ as an excellent, lovely, and rich instrument with a beautiful sound.

Out of around forty Hammond organs imported into Norway between 1936 and 1952, at least twenty ended up in Protestant churches, chapels, or crematoria, while one was delivered to a Catholic church.[43] One may ask why the church was so significant in the story of the Hammond organ in Norway. The lack of a lucrative home market during this early period is one explanation. Furthermore, most of Dahm's marketing was directed toward churches.

Majorstuen Church in Oslo (Figure 3.7) placed the first order with Arvid Dahm. In January 1937 the church acquired a Hammond Model BC with two B-40 tone cabinets. A few months later the priest and the organist expressed their satisfaction, and it was also their impression that the congregation shared their opinion.[44]

Holt Church in the small southern town of Tvedestrand was the second church to obtain a Hammond organ. A bank was the largest sponsor when a Model A with a B-40 tone cabinet was bought during the summer 1937. The nearly one-hundred-year-old pipe organ was still in place, and the congregation tried to make sure that the new

FIGURE 3.7
Majorstuen Church in Oslo (ca. 1935) was the first church in Norway to obtain a Hammond organ. Copyright © Wilse/Oslo Museum.

instrument did not interfere too much with the interior of the old church building. According to a local newspaper the expectations of the many people who showed up at the inauguration of the organ were fulfilled.[45]

Even in the northern and more sparsely populated parts of Norway the Hammond organ gained a foothold. Andenes Church was fortunate in 1938 to receive an organ donated by a local merchant (model unknown). The next year the organist proclaimed that "the Hammond organ is a wonder."[46] The church was without electricity and thus had to use a separate diesel generator to power the organ, although due to its noise, the Hammond organ was soon replaced by a pump organ.

Congregations, priests, and organists had several good reasons to favor the Hammond organ. A comparable pipe organ would cost many times as much. Another argument put forward was that the Hammond was not affected by the harsh Norwegian climate and could not get out of tune. Additionally, a three-year guarantee made it a rather safe investment. The small size was also a consideration. According to Orpha Ochse, Hammond and other producers of electronic organs gained "virtual control over the small-church market" in the United States after the war.[47] However, it is interesting to note that in Norway there were just as many large churches among those that acquired a Hammond organ. Both Majorstuen Church and Holt Church, for instance, were relatively large, with around seven hundred seats.

In some cases the Hammond organ was an alternative to the pump organ. Quite often the instrument was meant to be just a temporary investment until enough money could be raised to buy a new pipe organ or have the old one restored, yet it often turned out to be a permanent solution. Many of the Hammond organs acquired in the 1930s, 1940s, and 1950s were in use for twenty to thirty years, and a few were not replaced until around 1990.

A Hammond M2 spinet model used in Hopen Church in Smøla on the northwest coast of Norway may illustrate the fate of several organs. Hopen Church ordered the organ from Arvid Dahm in 1953. Initially the spinet models were intended for private homes, but since they were cheaper than the church models, some churches acquired such instruments instead. When the church took the organ out of service and advertised it for sale in 1980, no one showed any interest in buying, and finally it was given to a woman in exchange for a chasuble. Today it still has a place in her home (Figure 3.8).[48]

From Enthusiasm to Opposition

In 1951 Arvid Dahm asked Majorstuen Church for a statement concerning their experience with the Hammond organ after fifteen years of use. According to the organist and his colleagues the instrument had done its job as promised. In their opinion "the musical objections which it has now become modern to raise against the Hammond organ" were "highly exaggerated."[49] Nevertheless, after the war the enthusiasm of the 1930s increasingly turned into opposition toward the instrument. Many organists expressed their relief

FIGURE 3.8
Hammond M2 spinet model in the home of Ragnhild and Per Gjevik, Hopen, Smøla. Courtesy of Rannei Botten, Nordmøre Museum.

when, in the 1960s, Majorstuen Church replaced the "miserable" Hammond organ with a pipe organ.[50]

Another example may illustrate the differing opinions. The congregation of Våler Church in the southeast of Norway decided to buy a Hammond organ in 1941 (model unknown). The organ was installed during the spring of 1942 and was inaugurated on May 17 (the Norwegian National Day, which it was forbidden to celebrate during the German occupation). Since all imports from the United States were stopped during the war, the instrument was most likely imported by Dahm some years earlier. An organist described the new organ as "an excellent instrument for use in service, easy to operate and with a seldom-heard pure tone."[51] But there were many who became dissatisfied with the organ and accused it of being an unsuitable instrument, so in 1960 it was removed and sold. After the church had owned a Hammond organ for nearly twenty years the original pipe organ from 1868 was restored and reinstalled.[52]

The Norwegian Association of Organists gradually became a strong opponent of the Hammond organ. In 1937 the association's board made a rather moderate statement: "Despite the instrument's many good qualities and great possibilities of development, we find that the type which is now available is not yet so well experimented that it can fully replace a church organ for use in services."[53] Some years later the critical tone had become much sharper. At a national meeting in 1954 the association unanimously decided to raise its "sharpest protest against the installation of electronic instruments and other musically unsatisfying organ surrogates in Norwegian churches and chapels."[54] The protest was particularly addressed to "the so-called Hammond organ which uses speakers to reproduce the sound, while the organists wish a far more living sound from the pipe organs."[55]

During the 1950s the organists' association worked actively to inform organists, priests, congregations, and the public about its view of electronic organs. It criticized the musical policy of the Norwegian Broadcasting Corporation—which broadcast regular Hammond organ concerts—and called for a weekly radio concert with "real organ music."[56]

The association's policy changed little during the 1960s. In 1961 it requested the Bishops' Councils for Church Music to send strong warnings against the use of electronic organs to every congregation throughout the country. Not only were these instruments considered completely unsatisfactory and inappropriate, they were also considered to corrupt the understanding of real church music and thus could not even be used for practice.[57] In 1967 the association still found it "necessary to fight against the constantly increasing spread of electronic instruments in the churches."[58]

Not many organists publicly defended or spoke favorably of the Hammond organ after the war. Still, there were some who continued to send their recommendations to Arvid Dahm. Besides, we must assume that since many organists were part of church committees that chose to install Hammond organs, not all were as hostile to the new instruments as their association. In the organists' periodical in 1961 a vicar commented on what he regarded as the organists' "conservative one-sidedness." He argued that the Hammond organ was a new and independent instrument rather than a surrogate for something else. The article ended with a provoking pair of questions: "What if it develops into the instrument of the future? Who can guarantee this will not be the case?"[59]

To some extent the reactions against the Hammond organ might be understood in the light of a broader organ reform movement in Europe that enjoyed support in Norway after the Second World War. The movement was influenced by the famous theologian, physician, and organist Albert Schweitzer, who had expressed his dissatisfaction with the design and building of organs at the beginning of the twentieth century, and it was later formed by German organ builders. Adherents of the movement wanted to restore the dignity of the organ as an independent instrument and reestablish earlier tonal characteristics. The ideal was historically inspired baroque organs with mechanical key actions. In Norway these ideas were most strongly promoted by a society called Musica Sacra, established in 1952. One of the leading figures of the society, the organist Rolf Karlsen, related the so-called organ problem to the restoration of churches. Because of the small size of Norwegian medieval churches and the country's ancient wooden stave-churches, reed organs or electric organ surrogates were often chosen. According to Karlsen it was important to restore the dignity of the organ as well: "But there is something wrong in bringing the church back to its original, and then concerning the church music, settle for the unreal."[60] It should be noted, however, that when Karlsen referred to electric organ surrogates he seemed to be more concerned with electropneumatic pipe organs than with Hammond organs.

An Organ Surrogate

It is interesting to see how notable people within the Norwegian Association of Organists—not least Arild Sandvold, who was the chairman from 1935 to 1957—changed their view of the Hammond organ. When the instrument was introduced in the 1930s they described it in favorable terms, while after the war they clearly opposed it. Arvid Dahm's company continued to use the organists' recommendations in its advertisements after the war. This naturally provoked reactions from the organists, who also felt that their former statements were used in a distorted and misleading way.[61] But even though the hostility toward the Hammond organ in Norway became notably stronger after the war, the discussions concerning whether it could actually be called an organ had existed from the very beginning. Neither was the characterization of the instrument as a surrogate anything new.

The Hammond case in the United States in the 1930s was also debated in Norway. A newspaper reported in 1938 that the manufacturer of "the church organ surrogate known as the Hammond organ" was "sentenced to stop its false assertions." The company's "unfair competition" was "suited to harm the public by giving them false conceptions and had the tendency to betray a great part of the organ buyers by making them believe that the statements were true."[62] This implied that the company was no longer allowed to claim that the Hammond organ could produce the entire range of tone coloring necessary for the rendition of classical organ literature, or that it was comparable to a $10,000 pipe organ.

The newspaper article caused a reaction from Arvid Dahm. According to his lawyer the article was certainly inspired by pipe organ advocates. Newspaper jargon such as "misleading advertisements" or "false assertions" were not to be found in the verdict from the Federal Trade Commission. Furthermore, he argued that the crucial point had been whether the Hammond company "should still be allowed to call their instrument an *organ* and its music for organ music."[63] Since this had not been denied them, Hammond claimed victory.

However, the verdict from the Federal Trade Commission had limited influence on how others characterized the instrument. A Norwegian music encyclopedia published in 1949 grouped the Hammond organ among "organ imitating instruments" and typically defined the instrument by its lack of pipes.[64] The organists' association was even hesitant to use the word *organ*: the proper name was rather *Hammond elektroton* or the broader term *elektron instrument*.[65] This renaming can be seen as a parallel to the use of the term *electrotone* in the United States and Britain, and the attempts of the German Gesellschaft der Orgelfreunde to introduce the name *elektrium*.

The characterization "organ surrogate" was used both by the press and organists to describe the Hammond organ. Already in 1937, after the concert given by Arild Sandvold and the Oslo Philharmonic Society Orchestra, one journalist made a harsh statement: "That the Hammond organ is a surrogate is something everyone should know, but it is also a very poor surrogate."[66] When the Norwegian organists' association protested against the use of electronic organs, it even referred to such organs as "synthetic instrument surrogates."[67] The adjective *synthetic*, which had been commonly used by American critics since the 1930s, seems to have had a double meaning. On the one hand it referred to how sound was generated through additive synthesis; on the other hand it is likely to have implied that the new instruments were considered to be unnatural and artificial.

Another characterization pointed in the same direction. An article in the periodical of the Norwegian organists' association in 1961 described electronic organs as "electronic sound apparatuses."[68] The organs were seen as machinery that generated sound rather than living music. The fact that electronic organs were to a large degree machine made and mass produced probably contributed to these conceptions. As put by Walter Holtkamp, a well-known organ builder and one of the leading figures of the organ reform movement in the United States in the 1930s, the electronic organs were "machine-made, scientific substitutes."[69]

Laurens Hammond had claimed that while "the Hammond organ is played like a pipe organ, it is not made in imitation of one; it is a new musical instrument with a voice of its own."[70] Such statements were meant to counter the critique that the Hammond organ was just a surrogate for the pipe organ. Arvid Dahm could be quite aggressive when advertising the instrument. A "replacement of the old pipe organ with its many weaknesses" had been expected for some time, he claimed in 1939.[71] As one of his employees remembers, many organists found this attitude rather provocative.[72]

Conclusion: Technology and Authenticity

The British music sociologist Simon Frith has given some examples of controversies caused by the introduction of new music technology. In his essay "Art versus Technology: The Strange Case of Popular Music" he describes how crooning made possible by the electrical microphone was banned from the BBC in the 1930s; how Bob Dylan was met with fury when he toured Britain with his new electric band in 1966; and how a young band in Coventry in the 1980s was rejected at an event sponsored by the Musicians' Union because it used a drum machine. According to Frith, such examples suggest that technology is opposed to nature, community, and art—technology is assumed to be unnatural, alienating, and soul-less. "Authenticity" is a key concept in Frith's analysis:

"What is at stake in all these arguments is the authenticity or truth of music; the implication is that technology is somehow false or falsifying."[73]

In some sense the reception of the Hammond organ confirms the point made by Frith. Even though the instrument had its weaknesses compared with the pipe organ—and undoubtedly much of the critique from organists and others concerning the sound qualities was correct—the rejection of the instrument might suggest a view of new music technology as being unnatural and corrupting. The traditional pipe organs were of course also complex technological mechanisms, but still they were considered to be more worthy and soulful. The disputes about the term *organ* and the organists' efforts to prevent the use of the Hammond organ in churches can be seen as attempts to draw a sharp line between traditional and modern technology, between true musical instruments and false machines. Furthermore, the characterization of the Hammond organ as a surrogate emphasizes the notion of the instrument as something unauthentic. By its opponents it was seen as an illegitimate machine and an unwarranted intrusion into the established musical culture.

During the last decades, as the Hammond organ has gained a cult status within jazz, blues, rock, and popular music, conceptions of the instrument may be said to have changed in a rather ironic way. Hammond purists often describe the sound of the tonewheel organ as more alive, more organic, and warmer than the sound of modern electronic organs. The different Hammond models are also ascribed their own characteristics and personalities. In Mark Vail's book *The Hammond Organ: Beauty in the B*, which is devoted to the legendary B-3 tonewheel organ produced from 1955 to 1974, a Hammond enthusiast expresses his love for the real thing: "There's something very natural about the sound, almost like a violin or some other acoustic instrument, that makes it really pleasing."[74] Terms that were earlier used to distinguish pipe organs are now applied to the Hammond tonewheel organ. It has certainly become a musical instrument in its own right and with a voice of its own.

Today, digital musical instruments are being made to imitate the sound of the original tonewheel organ. Characteristically these are sometimes referred to as "clonewheel organs." While the Hammond tonewheel organ is now considered to be a genuine and authentic musical instrument, such digital instruments have become the subject of debate among musicians.

Notes

1. Trevor Pinch and Frank Trocco, *Analog Days. The Invention and Impact of the Moog Synthesizer* (Cambridge, MA, and London: Harvard University Press, 2002), 306-8. See also Trevor J. Pinch and Karin Bijsterveld, "'Should One Applaud?' Breaches and Boundaries in the Reception of New Technology in Music." *Technology and Culture* 44(3) (2003): 536–59.

2. Hammond Instrument Company, *The Hammond Organ* (Chicago, n.d.).

3. The archive at the Norwegian Museum of Science and Technology (hereafter NTM) includes incomplete lists of sales from the importer Dahm. Thanks to Tom Erik Ingvaldsen,

Wilfred W. Høsteland, Mats Krouthén, and Amanda Steggell. The model in Figure 3.2 and the Hammond BC organ in Figure 3.6 are owned by Ingvaldsen.

4. Barry Schrader, *Introduction to Electro-Acoustic Music* (Englewood Cliffs, NJ: Prentice-Hall, Inc., 1982), 68. Paul Théberge, *Any Sound You Can Imagine. Making Music/Consuming Technology* (Middletown, CT: Wesleyan University Press, 1997), 17, 41. Concerning the naming of such instruments and the terminological confusion around the labels *electric*, *electronic*, *electroacoustic*, and *electromechanical*, see Hugh Davies, "Electronic Instruments," in *The New Grove Dictionary of Musical Instruments* Vol. I, ed. S. Sadie (London: Macmillan Press Ltd., 1984), 658.

5. Hugh Davies, "Electronic Organ," in *The New Grove Dictionary of Musical Instruments* Vol I, ed. S. Sadie (London: Macmillan Press Ltd., 1984), 690. Hugh Davies, "Hammond Organ," in *The New Grove Dictionary of Musical Instruments* Vol II, ed. S. Sadie (London: Macmillan Press Ltd., 1984), 120. Mark Vail, *The Hammond Organ: Beauty in the B* (San Francisco: Backbeat Books, 2002), 38.

6. Peter Forrest, *The A–Z of Analogue Synthesisers. Part One: A–M* (Exeter, UK: Susurreal, 1998), 164.

7. George Yates Myers, "Has the Electric Organ a Place in Church Services?" *The Caecilia* 64(8) (1937): 359.

8. Hammond Instrument Company, *Hammond Organs* (Chicago, n.d.), 2.

9. Boosey & Hawkes Ltd., *The Hammond Organ* (London, n.d.), 3.

10. Hans-Joachim Braun, "Introduction," in *Music and Technology in the Twentieth Century*, ed. Hans-Joachim Braun (Baltimore, MD: The Johns Hopkins University Press, 2002), 12. Cf. Théberge, *Any Sound You Can Imagine*, 46.

11. *Hammond* Instrument Company, *Creating Beautiful Colors with the Harmonic Drawbars of the Hammond Organ* (Chicago, n.d.), 4.

12. Hammond Organ Company, *Fifty Years of Musical Excellence. Hammond Organ Company. 50th Anniversary 1934–1984* (Chicago, n.d.), 4.

13. Vail, *The Hammond Organ*, 36–37. Théberge, *Any Sound You Can Imagine*, 46.

14. Théberge, *Any Sound You Can Imagine*, 45.

15. Schrader, *Introduction to Electro-Acoustic Music*, 68.

16. Albert Glinsky, *Theremin. Ether Music and Espionage* (Chicago: University of Illinois Press, 2005). Frode Weium, "Ingenieurmusik. Die Begegnung mit elektrischen Musikinstrumenten in Norwegen in den 20er- und 30er-Jahren," *Blätter für Technikgeschichte* 69/70 (2008): 136.

17. Théberge, *Any Sound You Can Imagine*, 47.

18. Orpha Ochse, *The History of the Organ in the United States* (Indianapolis: Indiana University Press, 1988), 369.

19. Ibid., 371. Wilfred W. Høsteland, *The Historic and Famous Hammond Organ Church Test. Press Reports 1936–1937–1938* (Bergen, 2008).

20. Herbert Westerby, "Electronic Organs," *The Musical Times* 78(1128) (1937), 151. Stuyvesant Barry, *Hammond as in Organ. The Laurens Hammond Story*. http://thehammondorganstory.com (accessed December 23, 2009). Hammond Organ Company, *Fifty Years of Musical Excellence*, 7–9. Ochse, *The History of the Organ in the United States*, 371–73. Høsteland, *The Historic and Famous Hammond Organ Church Test.*

21. Anonymous, "Hammond 'Organ,'" *Catholic Choirmaster*, September 1949, 131, 142. Cf. Fiorenzo Romita, *La proibizione della S. Sede d'introdurre in chiesa il cosi detto organo "Hammond"*

(Rome: Tipografia Poliglotta Vaticana, 1949). Anonymous, "Hammond elektroton er forbudt," *Medlemsblad for Norges Organistforbund* 3 (March 1950), 26–27.

22. Vatican Council, *Constitution on the Sacred Liturgy. Sacrosanctum Concilium* VI (December 4, 1963). http://www.vatican.va/archive/hist_councils/ii_vatican_council/documents/vat-ii_const_19631204_sacrosanctum-concilium_en.html (accessed December 23, 2009), 120.

23. Anonymous, "Heilige Pfeifen," *Der Spiegel*, 16, 1972, 76–79. Some larger European cathedrals, such as those in Gnesen, Poland, and Split, Croatia, also had electronic organs.

24. Wolfgang Adelung, *Das Elektrium. Veröffentlichung der Gesellschaft der Orgelfreunde* Vol. 22 (Berlin: Verlag Merseburger, 1964). Anonymous, "Heilige Pfeifen." Davies, "Electronic Instruments," I, 658. Wolfgang Voigt, *Zum Problem der Akzeptanz und der "natürlichen" Klangwirkung elektronischer bzw. Mechanisch-elektronischer Musikinstrumente*. http://www.uni-koeln.de/phil-fak/muwi/fricke/307voigt.pdf (accessed December 23, 2009).

25. Arvid Dahm, "Hammondorgler levert i Norge," list of sales dated November 11, 1952, NTM. Regarding the company Dahm, see Høsteland, "Hammond's norske historie," *Hammond Retrospect* 3 (2006), 1–7. In 1950 Dahm asked the Norwegian Ministry of Church and Education to recommend an application for funding necessary to start building Hammond organs on license. No answer has been found, but most likely this request was turned down. Cf. Kirke- og undervisningsdepartementet to Norges Organistforbund, February 24, 1951, The National Archives of Norway, Jnr. 6623-A-1950.

26. Sandvold is quoted in Anonymous, "Elektrisk orgel," *Morgenposten*, February 9, 1937, 3.

27. Anonymous, "Hammond-orglet som har 253 millioner forskjellige toner," *Dagbladet*, February 9, 1937, 15.

28. Anonymous, "Hammond-orglet," *Norges Handels- og Sjøfartstidende*, February 9, 1937, 2.

29. Ulrik Mørk, "Filharmoniske," *Nationen*, November 16, 1937, 3.

30. Thorleif Eken, "Filharmonisk konsert," *Morgenposten*, November 16, 1937, 2.

31. Anonymous, "Hammond-orgelet," *Tonekunst* 18, 1937, 159. L. Saxegaard, "Hammond-Orglet," *Tonekunst* 19, 1937, 171.

32. Anonymous, "Oslo Philharmonic Orchestra," *Hammond Times*, Vol. XIV, 10 (1952): 5.

33. According to the Social Construction of Technology (SCOT), the development of a technological artifact is negotiated between relevant social groups such as users, producers, politicians, and others. Trevor Pinch and Frank Trocco have studied Bob Moog's synthesizer from this perspective. Pinch and Trocco, *Analog Days*.

34. Anonymous, "Elektromusikk," *Teknisk Ukeblad*, May 6, 1937, 227.

35. Georg Brochmann, *Mennesket og maskinen* Vol II (Oslo: Aschehoug, 1937), 154.

36. Kjell Mørk Karlsen, e-mail to author, August 24, 2008.

37. Gert Prix, "The Devil's Work or a Musical Instrument to be Taken Seriously?" in IMA Institut für Medienarchäologie, *Zauberhafte Klangmaschinen. Von der Sprechmaschine bis zur Soundkarte* (Mainz: Schott Music, 2008), 163.

38. Production of the Hammond Model H-112 started in 1965. As seen in Figure 3.5, the original preset keys and drawbars have been removed. The BC model acquired by the Norwegian Broadcasting Corporation in 1937 was refurbished in a similar way. Hakon Blandehoel, message to author, September 14, 2010.

39. Hakon Blandehoel, message to author, November 16, 2009.

40. Private recordings with Einar Iversen (1957) and Kjell Karlsen (ca. 1959). Bjørn Stendahl, e-mail to author, August 25, 2011.

41. Around 1950 a Hammond Model CV organ was also acquired by the new City Hall in Oslo.

42. Arvid Dahm, *Hammond orgel-nytt*, No. 2 (1939), 2. Copies of numerous letters from organists to Arvid Dahm written between 1937 and 1951 are to be found in the NTM archive.

43. See Dahm, "Hammondorgler levert i Norge." In the United States the relative proportion of organs sold to churches was lower from the start. During the first three years of manufacture some 1,750 churches in the United States purchased a Hammond organ, which represented about a third of all sales (Davies, "Hammond Organ," 121).

44. Carsten Lund Iversen to Arvid Dahm, June 14, 1937 (NTM). Johan Prytz to Arvid Dahm, September 18, 1937 (NTM).

45. Anonymous, "Holt kirkes nye orgel," *Agderposten*, September 2, 1937, 3. Arvid Dahm, *The Hammond Organ* (Oslo, 1937).

46. Helmer Nordgaard to Arvid Dahm, March 21, 1939 (NTM). Andreas Blix, message to author, August 27, 2008.

47. Ochse, *The History of the Organ in the United States*, 374.

48. Odd Williamsen, e-mail to author, January 21, 2009.

49. Carsten Lund Iversen, Johan Prytz, and Karl Hafstad to Arvid Dahm, September 15, 1951 (NTM).

50. Kjell Mørk Karlsen, e-mail to author, August 24, 2008. A. Sandvold, "Oslo's orgler," in *Oslo Organistforening 50 år* (Oslo: A.S. Industritrykkeriet, 1966), 33.

51. Knut Nystedt to Arvid Dahm, May 8, 1937 (NTM).

52. Ivar J. Hauge, *Våler kirke gjennom 800 år* (Våler, Norway: Våler kommune, 1998), 25.

53. Arild Sandvold, Oskar Skaug, and Trygve Præsttun, *Norges organistforening. Beretning gjeldende årene 1936–46* (The Labour Movement Archive and Library in Norway. Norsk kantor- og organistforbund, 2567—hereafter ARBARK-2567).

54. Erling Kjelsen, "NOF.s landsmøte i Trondheim 31. mai 1954," *Medlemsblad for Norges Organistforbund* 6 (June 1954), 55.

55. Anonymous, "Ikke Hammond-orgel i kirkene," *Verdens Gang*, June 1, 1954, 10. H. Gaare, "Voksende forståelse for kirkemusikken," *Verdens Gang*, June 3, 1954, 3.

56. O. R. Krag, "Rikskringkastingens positiv," *Medlemsblad for Norges Organistforbund* 5 (1951), 37. Ludvig Nielsen, Rolf Karlsen, and Per Hjort Albertsen, *Musikkutvalget: Forberedende møte 19/8 1957* (ARBARK-2567). Ludvig Nielsen, Rolf Karlsen, and Per Hjort Albertsen, *Musikkutvalget: Beretning om utvalgets arbeid 1956–58* (ARBARK-2567).

57. Egil Hovland, "Hammond-instrumentet," *Norsk Kirkemusikk* 7 (1961), 121. Egil Hovland, "Elektroninstrumenter," *Norsk Kirkemusikk* 10 (1961), 179. Anonymous, "Elektronorgler er ikke egnet til kirkelig bruk," *Norsk Kirkemusikk* 2 (1962), 33. Anonymous, "Orgel-debatt," *Norsk Kirkemusikk* 1 (1961), 3.

58. C. B. "Organistene frarår elektroninstrumenter," *Vårt Land*, July 3, 1967, 2.

59. Sverre Barstad, "Hammond-instrumentet," *Norsk Kirkemusikk* 7 (1961): 120.

60. Rolf Karlsen, "Orgelproblemer," in *Foreningen til norske fortidsminnesmerkers bevaring, Årbok 1953* (Oslo: Fortidsminneforeningen, 1953), 46.

61. Kjelsen, "NOF.s landsmøte i Trondheim 31. mai 1954," 55.

62. Anonymous, "En amerikansk orgelprosess," *Morgenbladet*, August 27, 1938, 5.

63. E. W. Nansen, "En amerikansk orgelprosess," *Morgenbladet*, September 6, 1938, 9.

64. Olaf Gurvin and Øyvind Anker, *Musikkleksikon* (Oslo: Dreyers Forlag, 1949), 280, 438.

65. Anonymous, "Hammond elektroton er forbudt," 26–27. Cf. Sandvold, "Oslo's orgler," 33.

66. Eken, "Filharmonisk konsert," 2.

67. Anonymous, "Ikke Hammond-orgel i kirkene," 10.

68. Egil Hovland, "Hva nå?" *Norsk Kirkemusikk* 5 (1961): 75.

69. Walter Holtkamp, "Organ Builder Calls for Introspection by His Profession," *The Diapason* 27(3) (1936): 24.

70. Barry, *Hammond as in Organ.*

71. Dahm, *Hammond orgel-nytt*, No. 2 (1939): 2.

72. Hakon Blandehoel, message to author, November 16, 2009.

73. Simon Frith, "Art versus Technology: The Strange Case of Popular Music," *Media, Culture & Society* 8 (1986): 265.

74. Vail, *The Hammond Organ*, 10.

Bibliography

Adelung, Wolfgang. *Das Elektrium. Veröffentlichung der Gesellschaft der Orgelfreunde*. Vol. 22. Berlin: Verlag Merseburger, 1964.

Anonymous. "Elektrisk orgel." *Morgenposten*, February 9, 1937, 3.

———. "Elektromusikk." *Teknisk Ukeblad*, May 6, 1937, 227.

———. "Elektronorgler er ikke egnet til kirkelig bruk." *Norsk Kirkemusikk* 2 (1962): 33.

———. "En amerikansk orgelprosess." *Morgenbladet*, August 27, 1938, 5.

———. "Hammond elektroton er forbudt." *Medlemsblad for Norges Organistforbund* 3 (March 1950): 26–27.

———. "Hammond 'Organ.'" *Catholic Choirmaster*, September 1949, 131, 142.

———. "Hammond-orgelet." *Tonekunst*, 18, 1937, 159.

———. "Hammond-orglet." *Norges Handels- og Sjøfartstidende*, February 9, 1937, 2.

———. "Hammond-orglet som har 253 millioner forskjellige toner." *Dagbladet*, February 9, 1937, 15.

———. "Heilige Pfeifen." *Der Spiegel*, 16, 1972, 76–79.

———. "Holt kirkes nye orgel." *Agderposten*, September 2, 1937, 3.

———. "Ikke Hammond-orgel i kirkene." *Verdens Gang*, June 1, 1954, 10.

———. "Orgel-debatt." *Norsk Kirkemusikk* 1 (1961): 3.

———. "Oslo Philharmonic Orchestra." *Hammond Times*, Vol. XIV, 10 (1952): 5.

Barry, Stuyvesant. *Hammond as in Organ: The Laurens Hammond Story.* http://thehammondorganstory.com (accessed December 23, 2009).

Barstad, Sverre. "Hammond-instrumentet." *Norsk Kirkemusikk* 7 (1961): 119–20.

Boosey & Hawkes, Ltd. *The Hammond Organ*. London, n.d.

Braun, Hans-Joachim. "Introduction." In *Music and Technology in the Twentieth Century*, pp. 9–32, ed. Hans-Joachim Braun. Baltimore, MD: The Johns Hopkins University Press, 2002.

Brochmann, Georg. *Mennesket og maskinen*, Vol. II. Oslo: Aschehoug, 1937.

C. B. "Organistene frarår elektroninstrumenter." *Vårt Land*, July 3, 1967, 2.

Chadabe, Joel. *Electric Sound: The Past and Promise of Electronic Music*. Upper Saddle River, NJ: Prentice Hall, 1997.

Dahm, Arvid. *The Hammond Organ*. Oslo, 1937.

———. *Hammond orgel-nytt*, No. 2, 1939.

———. "Hammondorgler levert i Norge." List of sales dated November 11, 1952. The Norwegian Museum of Science and Technology (hereafter NTM).

Davies, Hugh. "Electronic Instruments." In *The New Grove Dictionary of Musical Instruments*, Vol. I, p. 658, ed. S. Sadie. London: Macmillan Press Ltd., 1984.

———. "Electronic Organ." In *The New Grove Dictionary of Musical Instruments*, Vol. I, pp. 690–91, ed. S. Sadie. London: Macmillan Press Ltd., 1984.

———. "Hammond Organ." In *The New Grove Dictionary of Musical Instruments*, Vol. II, pp. 120–22, ed. S. Sadie. London: Macmillan Press Ltd., 1984.

Eken, Thorleif. "Filharmonisk konsert." *Morgenposten*, November 16, 1937, 2.

Forrest, Peter. *The A–Z of Analogue Synthesisers. Part One: A–M.* Exeter: Susurreal, 1998.

Frith, Simon. "Art versus Technology: The Strange Case of Popular Music." *Media, Culture & Society* 8 (1986): 263–79.

Gaare, H. "Voksende forståelse for kirkemusikken." *Verdens Gang*, June 3, 1954, 3.

Glinsky, Albert. *Theremin. Ether Music and Espionage*. Chicago: University of Illinois Press, 2005.

Gurvin, Olaf, and Øyvind Anker. *Musikkleksikon*. Oslo: Dreyers Forlag, 1949.

Hammond Instrument Company. *Creating Beautiful Colors with the Harmonic Drawbars of the Hammond Organ*. Chicago, n.d.

———. *The Hammond Organ*. Chicago, n.d.

———. *Hammond Organs*. Chicago, n.d.

———. *Music That Makes a Happier Home*. Chicago, n.d.

Hammond Organ Company. *Fifty Years of Musical Excellence. Hammond Organ Company. 50th Anniversary 1934–1984*. Chicago, n.d.

Hauge, Ivar J. *Våler kirke gjennom 800 år*. Våler, Norway: Våler kommune, 1998.

Holtkamp, Walter. "Organ Builder Calls for Introspection by His Profession." *The Diapason* 27(3) (1936): 24.

Høsteland, Wilfred W. "Hammond's norske historie." *Hammond Retrospect* 3 (2006): 1–7.

———. *The Historic and Famous Hammond Organ Church Test. Press Reports 1936–1937–1938*. Bergen, 2008.

Hovland, Egil. "Elektroninstrumenter." *Norsk Kirkemusikk* 10 (1961): 179.

———. "Hammond-instrumentet." *Norsk Kirkemusikk* 7 (1961): 120–21.

———. "Hva nå?" *Norsk Kirkemusikk* 5 (1961): 75.

Iversen, Carsten Lund to Arvid Dahm, June 14, 1937. NTM.

Iversen, Carsten Lund, Johan Prytz, and Karl Hafstad to Arvid Dahm, September 15, 1951. NTM.

Karlsen, Rolf. "Orgelproblemer." *Foreningen til norske fortidsminnesmerkers bevaring, Årbok 1953*. Oslo: Fortidsminneforeningen, 1953, 35–46.

Kirke- og undervisningsdepartementet to Norges Organistforbund, February 24, 1951. The National Archives of Norway, Jnr. 6623-A-1950.

Kjelsen, Erling. "NOF.s landsmøte i Trondheim 31. mai 1954." *Medlemsblad for Norges Organistforbund* 6 (June 1954): 55.

Krag, O. R. "Rikskringkastingens positiv." *Medlemsblad for Norges Organistforbund* 5 (1951): 37–38.

Mørk, Ulrik. "Filharmoniske." *Nationen*, November 16, 1937, 3.

Myers, George Yates. "Has the Electric Organ a Place in Church Services?" *The Caecilia* 64(8) (1937): 359–60.

Nansen, E. W. "En amerikansk orgelprosess." *Morgenbladet*, September 6, 1938, 9.

Nielsen, Ludvig, Rolf Karlsen, and Per Hjort Albertsen. *Musikkutvalget: Beretning om utvalgets arbeid 1956–58*. The Labour Movement Archive and Library in Norway. Norsk kantor- og organistforbund, 2567 (hereafter ARBARK-2567).

———. *Musikkutvalget: Forberedende møte 19/8 1957.* (ARBARK-2567).

Nordgaard, Helmer to Arvid Dahm. March 21, 1939. NTM.

Nystedt, Knut to Arvid Dahm, May 8, 1937. NTM.

Ochse, Orpha. *The History of the Organ in the United States*. Indianapolis: Indiana University Press, 1988.

Pinch, Trevor J., and Karin Bijsterveld. "'Should One Applaud?' Breaches and Boundaries in the Reception of New Technology in Music." *Technology and Culture* 44(3) (2003): 536–59.

Pinch, Trevor, and Frank Trocco. *Analog Days. The Invention and Impact of the Moog Synthesizer*. Cambridge, MA: Harvard University Press, 2002.

Prix, Gert. "The Devil's Work or a Musical Instrument to be Taken Seriously?" In IMA Institut für Medienarchäologie, *Zauberhafte Klangmaschinen. Von der Sprechmaschine bis zur Soundkarte*. Mainz: Schott Music, 2008.

Prytz, Johan to Arvid Dahm, September 18, 1937. NTM.

Romita, Fiorenzo. *La proibizione della S. Sede d'introdurre in chiesa il cosi detto organo "Hammond."* Rome: Tipografia Poliglotta Vaticana, 1949.

Sandvold, Arild. "Oslo's orgler." In *Oslo Organistforening 50 år*, pp. 25–34. Oslo: A. S. Industritrykkeriet, 1966.

Sandvold, Arild, Oskar Skaug, and Trygve Præsttun. *Norges organistforening. Beretning gjeldende årene 1936–46*. ARBARK-2567.

Saxegaard, L. "Hammond-Orglet." *Tonekunst* 19, 1937.

Schrader, Barry. *Introduction to Electro-Acoustic Music*. Englewood Cliffs, NJ: Prentice-Hall Inc., 1982.

Théberge, Paul. *Any Sound You Can Imagine. Making Music/Consuming Technology*. Middletown, CT: Wesleyan University Press, 1997.

Vail, Mark. *The Hammond Organ: Beauty in the B*. San Francisco: Backbeat Books, 2002.

Vatican Council. *Constitution on the Sacred Liturgy. Sacrosanctum Concilium*. December 4, 1963. http://www.vatican.va/archive/hist_councils/ii_vatican_council/documents/vat-ii_const_19631204_sacrosanctum-concilium_en.html (accessed December 23, 2009).

Voigt, Wolfgang. *Zum Problem der Akzeptanz und der "natürlichen" Klangwirkung elektronischer bzw. Mechanisch-elektronischer Musikinstrumente*. http://www.uni-koeln.de/phil-fak/muwi/fricke/307voigt.pdf (accessed December 23, 2009).

Weium, Frode. "Ingenieurmusik. Die Begegnung mit elektrischen Musikinstrumenten in Norwegen in den 20er- und 30er-Jahren." *Blätter für Technikgeschichte* 69/70 (2008): 135–54.

Westerby, Herbert. "Electronic Organs." *The Musical Times* 78(1128) (1937): 151–52.

CHAPTER 4

Mimics, Menaces, or New Musical Horizons?

Musicians' Attitudes toward the First Commercial Drum Machines and Samplers

Sarah Angliss

THE FIRST electronic drum machines and samplers presented musicians with many new ways to create and perform music. With a drum machine, a dance band could play polyrhythms, for the first time ever performing at lightning tempi without missing a beat. And with a sampler, musicians could slice up and recombine recorded sounds with ease, making their own musical collages. Despite their attractions, the drum machine and sampler could also be viewed as unwelcome cases of mechanization in the workplace. With these machines, anyone with an ounce of skill could create music (at least, according to the advertisers.[1,2]) Experienced musicians, it was averred, would no longer be in demand. A drum machine could generate a passable rhythm section at the touch of a button. And with a sampler, a keyboardist could play a convincing replica of almost any instrument—and even command a "whole orchestra"—with "just a thumb and two fingers." The drum machine and sampler loosened the coupling between musical sounds and the acoustic instrumentalists who made them. In this way, they continued a trend that began in 1877 with the invention of the phonograph.

Drum machines first appeared in mainstream music shops and magazines in 1959. By then samplers had already arrived, in analog form, and had been on sale for three years. This chapter shows how musicians weighed up the pros and cons of these machines as they decided how to exploit them (and whether they should use them at all). It also shows how attitudes to these devices changed over time. With the benefit of hindsight, it outlines the strengths and weakness of their arguments against these machines.

Some of the fiercest criticisms of the drum machine came from the United Kingdom's Musicians' Union, an organization that originated centuries earlier to protect musicians' jobs and welfare.[3] The drum machine was the first of these instruments to come under their scrutiny. In criticizing the drum machine, the Musicians' Union found itself in difficult philosophical territory. On the one hand, the union treated it just like any other capital that might be brought into the workplace. Drum machines could compete with "underdeveloped" drummers and rob them of opportunities to improve on the job.[4] On the other hand, the union argued that music was a creative artifact. It could not be judged on the same terms as a washing machine, electric oven, or other product that could be made cheaper—and therefore more desirable—using machines.[5] Interestingly, in criticizing the drum machine, the Musicians' Union revisited arguments they had already used against film sound half a century earlier. And although they mounted a vociferous attack on the drum machine, they spared the first sampler their opprobrium.[6]

The union was right to think that in the field of music, aesthetic concerns carry considerable weight and can often usurp economic judgments. In fact, as I explore in more detail below, their campaign against "canned music" failed because cinema audiences were intoxicated by the new worlds brought to them by sound film, regardless of any sympathies for the cinema orchestra who were put out of work. But the Musicians' Union failed to predict how far aesthetic judgments would shift, as mechanization made its mark on the music industry and the wider world. In particular, they didn't foresee how musicians in the late 1970s and early 1980s would find a new musical aesthetic that relished alienation, roboticism, and a lack of expressive musical dynamics. This style developed, in Europe (with bands such as Kraftwerk and Neu!) and in America (with movements such as Afrofuturism and Detroit techno), as a heartfelt response to people's mechanized, postindustrial landscapes. The drum machine and sampler offered the perfect machine embodiment of this aesthetic, as did the synthesizer. The unwavering tempo of the drum machine made it more appropriate for this kind of music than a human drummer, rather than a poor substitute for the real thing. And the sampler made it easier than ever to decontextualize, slice up, repitch, and recombine sounds. It realized the aspirations of both the European experimental music scene and the New York block party DJs. The latter had been splicing and remixing twentieth-century music culture on their turntables at the dawn of the hip-hop era.

Today, it's hard to find a single track on the popular music charts that has been made without a drum machine or sampler (in hardware or software form). This acceptance should come as no surprise—musicians have always been consummate cyborgs, teaming their minds and bodies with the latest technology to push their performance envelope. The drum machine and sampler can be viewed as the latest in a long line of instruments that have augmented the capabilities of human players—a lineage that can be traced back to the earliest keyboards, strings, flutes, and drums.

At the end of this chapter, I ask how much these factors have conspired to transform notions of musical authenticity and ownership. I argue that these terms have proved to be far more flexible than the Musicians' Union expected. But they are notions that continue to trouble musicians today, in the era of beat matching, Antares Auto-tune (pitch-correction software), and MP3 downloading. Examining these concerns, I also give an example of a contemporary musician who feels musically compromised if he's asked to perform with a human, rather than with his beloved samples and drum machines.[7]

This chapter focuses on three of the earliest drum machines: the Wurlitzer SideMan, Rhythm Ace, and Maestro Rhythm King; the first two samplers: the Chamberlin and Mellotron, both of which used magnetic tape, and the Fairlight Computer Music Instrument, a digital music sampler that followed them. It also looks at the contribution of the first turntablists (virtuosic DJs) to the art of sampling.[8]

The First Drum Machine

Between articles on a gas-powered fire engine and a metal lathe programmed using punched paper tape, the November 1960 edition of *Popular Mechanics* printed news of a musical invention.[9] The machine in question was the Wurlitzer SideMan, a machine that played drumlike electronic clicks and pops in never-ending rhythmic patterns. This assembly of electronic and mechanical parts was housed in a varnished wooden box around the size of a bedside cabinet (Figure 4.1, top).

In musical circles, a sideman is a skilled musician who can be hired for one-off gigs when extra players are needed. "Now even that job is being taken over by electronic substitutes," the article explained. "The new Wurlitzer 'Sideman' produces the effects of a rhythm section automatically . . . in 85 different beat patterns and at any speed from 36 to 195 beats per minute."

In May of that year, the SideMan was advertised in *Billboard*, an American music magazine. "Now you can be a combo all by yourself!" claimed the advertisement, which showed a smiling organ player adjusting the SideMan with one hand as he held down a chord with the other. "Flip a switch . . . and there's a rhythm section at your side! . . . Whether you play the piano, organ or any other instrument, professionally, or for your own enjoyment, the Wurlitzer Side Man provides the perfect rhythm accompaniment."[10]

For $395 (about the cost of a refrigerator) the SideMan really could turn any player into a fashionable music and rhythm combo, supplying regimented electronic beats in waltz, march, rhumba, tango, and other rhythms at the turn of a dial. This was the first drum machine from Wurlitzer, a company who was better known for their organs and jukeboxes. Arguably, it was also the first synthetic drum machine to go on sale. Earlier automatic drummers, such as the Rhythmicon (1931) from Henry Cowell and Leon Theremin, and Raymond Scott's Circle Machine (1959), were one-off experimental units. The Chamberlin Rhythmate had been on sale since 1948, but only a dozen or so were sold. Using drum sounds prerecorded on tape loops, rather than electronically synthesized beats, the Rhythmate was a precursor of the Mellotron—more about this later.

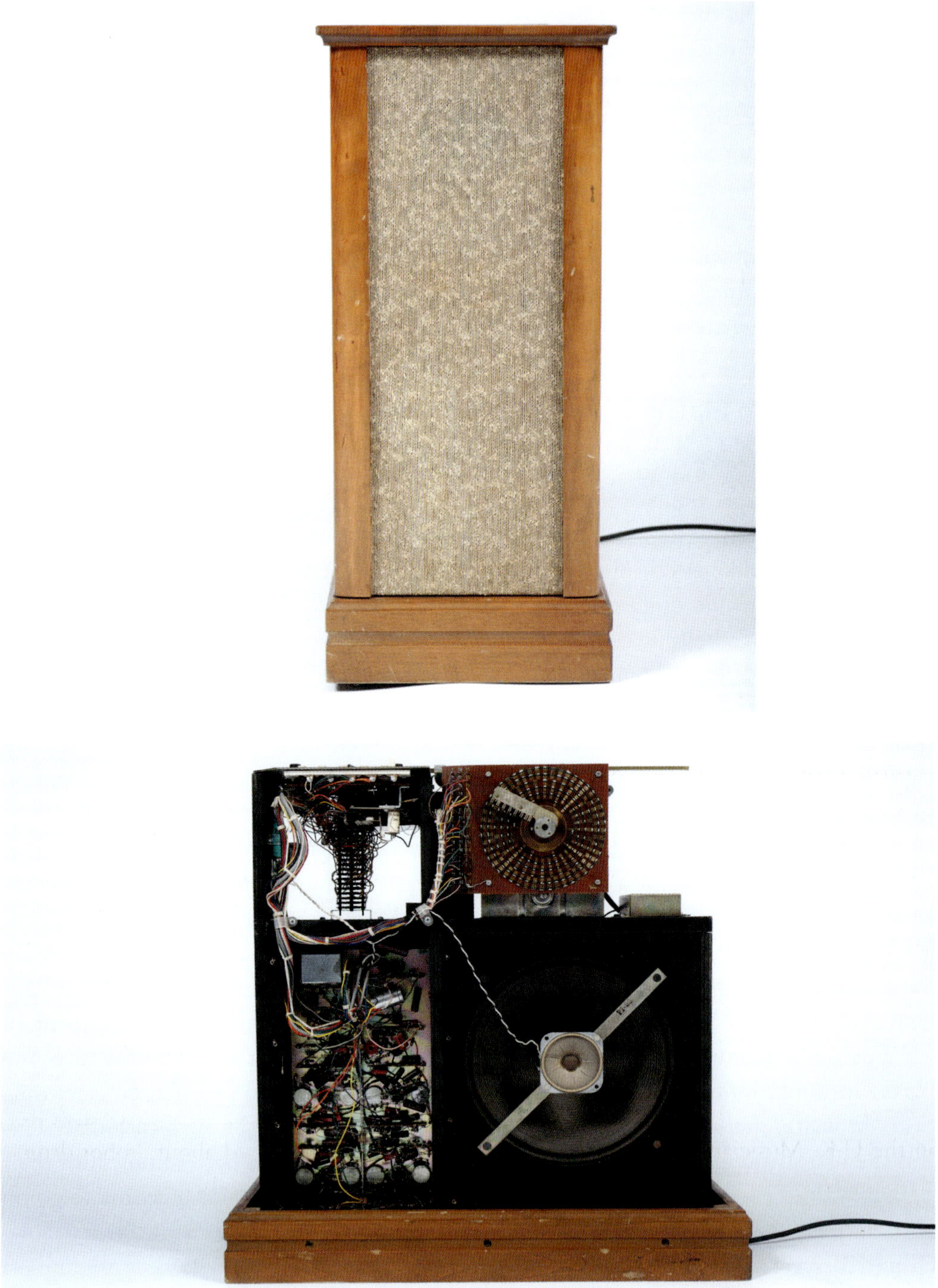

FIGURE 4.1
A Wurlitzer SideMan from 1961. Photo by Don Kennedy. Courtesy of National Music Centre, Canada (top). Inside the Wurlitzer SideMan, showing the rotor arm and hardwired sequencer. Photo by Don Kennedy. Courtesy of National Music Centre, Canada (bottom).

Under the Lid

Like other Wurlitzer products from the late 1950s, the SideMan combines valve electronics with motorized mechanical parts. Inside there is a hardwired sequencer, in the form of electrical contacts, which are arranged on a circuit board in several concentric circles (Figure 4.1, bottom). A rotor arm, with sprung metal teeth, sweeps around this board continually. As the arm rotates, each metal tooth momentarily brushes over a contact, triggering a pulse that energizes one of the SideMan's sound-making circuits. Some of these circuits oscillate at regular frequencies (for sonorous, bass drum notes); others are filtered noise generators (for cymbal crashes).

The SideMan makes ten types of percussion sound, including bass drum, cymbal, wood block, claves, tom-tom, and maracas. Dance styles are selected by turning a large dial that is itself attached to a rotary switch. This channels electrical pulses to the sound-making circuits required to make the chosen rhythm pattern.

Signals play through the SideMan's built-in speakers, which include a twelve-inch subwoofer for rich bass sounds, and two tweeters for the treble of the clicks and cymbal crashes. One tweeter is mounted just under the control panel, on the top of the SideMan, so users can monitor sounds while sitting at the keyboard.

The rotor arm is attached to a large circular plate, which is turned by a small idler wheel. This is coupled to the SideMan's motor by a drive belt that is made of stretched rubber to keep everything in tension. The SideMan can play at varying speeds (measured in beats per minute). Its speed control is simply a rod that pushes the idler wheel nearer or further from the arm's center of rotation. When the idler wheel moves toward the center, the arm takes less time to complete a rotation, so the machine plays faster rhythms.

The SideMan's volume and speed control knobs are chunky enough to use during a live performance. There are also trim pots that you can turn with a screwdriver to vary the tone of the instrument. These adjust the resonant frequency of the drum circuits or the decay rate of the "shimmer generator," which creates the long tail of the cymbal crash. There are also buttons that play a single shot of each percussion sound when they are pressed momentarily. Otherwise, the SideMan plays a new pattern and changes speed, volume, or tone only when directed to do so by a human operator. It endlessly repeats the same rhythm with metronomic precision, insensate to any other music that's playing. There's none of the *rubato* or other fine dynamic variation you'd get from a real drummer. This fact wasn't lost on the U.K. Musicians' Union, who became increasingly concerned about "robot drummers" such as the SideMan, which were proliferating in the early 1960s.

The Dilemma

In naming it the SideMan, Wurlitzer was pitching their musical instrument as an economic proposition. And in pointing up the convenience of a "rhythm section at your side," they were glossing over its obvious deficiency—the loss of a human drummer who can interact with other live musicians.

The SideMan was an expensive item and could hardly be described as portable (it weighed thirty-four kilograms). But it inspired no end of cheaper, lighter alternatives.[11] Two notable early machines are the Donca Matic (1962) and R1 Rhythm Ace (1964), both made by Japanese inventors who tried the SideMan and wanted to improve on it. The R1 Rhythm Ace played only single-shot sounds, but a second-generation model, the FR1, also played rhythm patterns. The Rhythm Ace drum machines were the first to abandon valves and mechanical sequencers altogether, creating sounds and measuring out rhythms using transistorized, electronic circuits. This move to solid-state electronics made them lighter and more reliable than their electromechanical predecessors. The makers of Donca Matic and the Rhythm Ace became Korg and Roland respectively. Their early machines inspired many copycats, including the Maestro Rhythm King.

By the end of the 1960s, drum machines really were cheap and light enough to offer a practical alternative to a human drummer—a boon for keyboard players who could afford one. With them, they could play percussive dance music, without needing to share the stage (and split the fee) with another player. Of course, this wasn't such good news for drummers. Arguably, they were in the same economic jeopardy as any other skilled laborer in competition with a machine, such as a metalworker about to be ousted by a new automatic lathe.

As musicians were encountering their first electronic drum machines, workers in industries across the world, from the car plants of Detroit to the foundries of Rotherham, found their jobs were being transformed by automatic conveyors, lathes, cutting machines, and other mechanized replacements for human labor. New forms of mechanization, which appeared in the 1960s, merely continued a trend that began in the late eighteenth century, when factory owners invested in powered machines that could outperform skilled human laborers.[12] In the 1950s and 1960s, as electronics enabled machines to run ever more complex routines, such as cutting shapes to order, ever more skilled workers found their expertise had been rendered obsolete by these new means of production. Was there any reason to think the music industry would be untouched by such changes? And how did professional musicians fit into this scenario? Were they like factory owners, weighing up the optimum mix of capital and human labor to create a desired product (in this case, music) and maximize profits? Or were they skilled artisans in direct competition with machines, like the weavers of the eighteenth century?[13]

The Musicians' Union was in no doubt that musicians were in competition with machines. After viewing a drum machine at an audio fair in 1967, the union outlined their concerns in an article titled "Don't Fall for It":

> When used automatically, these devices produce an accurate though rather stilted and unimaginative rhythm, but they can nevertheless be used by solo pianists, organists and others who want to curry favour with their employers by giving that "little bit extra" at no extra charge . . . It is the non-manual use of the robot "drummer" that is the biggest danger to the profession . . . when you are told that the electronic or robot

> "drummer" is merely a novelty, something to be lightly dismissed, don't fall for it! . . . there are as yet underdeveloped lesser known performers who could not stand the competition of the robot. But, if they are now put out of work by the robot, they are never going to have a chance to improve their standard or gain acceptance.[14]

At the end of the article, union members were instructed to boycott the robot drummer to protect the future of live music. In describing the drum machine as a "stilted" and "unimaginative" rhythm generator, one that can only compete with the most "underdeveloped" human player, the Musicians' Union was playing their trump card. According to their argument, music cannot be viewed like other artifacts, as it requires human expression. And the "robot drummer" is a poor substitute for a live musician because its performance has none of the dynamic qualities of an expressive human player. Quotation marks around the word "drummer" remind us that this electromechanical box of tricks shouldn't be thought of as a real performer at all. Note how little the union's critique said about the quality of the machine's synthetic drum sounds (its bass drum kicks, cymbal crashes, and so on). Expression evidently mattered more than realism. This line of argument wasn't new to the Musicians' Union, whose purpose had always been to protect the welfare of its members.

Machines and the Musicians' Union

The Musicians' Union originated in the trades guilds, which had existed since the fourteenth century to control the rights of musicians to perform in public.[15] It is also rooted in the mutual aid societies that appeared in the eighteenth century to provide benefits for part-time and provincial musicians. Over the next century, an assortment of guilds and societies flourished, competed, merged, radicalized, or folded, eventually forming the Amalgamated Musicians' Union in 1893, essentially the union we know today. Its first successful battle was against a five-shilling pay cut at the Carl Rosa's Court Theatre, Liverpool. Here, orchestra members also pooled their resources to pay the wages of the horn section that was laid off during the dispute.

At the turn of the twentieth century, most union business concerned wages, welfare, and the fair distribution of work. Musicians largely saw themselves in competition with each other, rather than with machines.[16] The union sometimes had to adjudicate between rival players, for instance between civilian musicians and the policemen and military bandsmen who undercut their rates. Similar internal battles were being fought in North America. In 1888, civilian players in New York noted one hundred instances in which military bandsmen, subsidized by their main paymasters (taxpayers), had been employed instead of civilian musicians, pushing down salaries well below rates agreed by their union, the National League of Musicians.[17] But in general, the early twentieth century was a boom time for live musicians. As recorded sound was still in its infancy, most people still expected music in theaters, ballrooms, and other public venues to be delivered by live musicians. And the cinemas, which were opening in every town, regularly employed their own pianists, organists, or bands to accompany silent moving pictures. By 1924, the Musicians' Union had well over twenty thousand members, half of whom regularly worked in cinemas.[18] Similarly, in North America, a quarter of all live musicians in 1928 worked in theaters and cinemas, and a musician could earn twice the weekly wage of a skilled worker in the building trade.[19] Mass-produced clockwork music boxes, such as the Polyphon and Symphonion, were used in some public spaces to play tunes automatically. But their metallic sounds were nothing like those of an organ, piano, violin, or other popular orchestral instruments. Public recitals by reproducing pianos were not unheard of, but these were still a novelty. Confronted with a machine replaying a human performance, encoded as perforations on rolls of paper, audiences in 1900 wondered "should one applaud"?[20]

Of course, the Musicians' Union didn't speak for everyone in the music supply chain. It didn't represent the interests of amateur players, for example, and it was often in conflict with the owners of studios, cinemas, dance halls, and other venues that hired its members for work. Vigilant for any development that might threaten members' interests, the Musicians' Union may have been quick to emphasize the drawbacks of certain innovations rather than relish any new musical possibilities on offer. So musicians interested in experimenting with new machines may have found themselves in conflict with the union's official line.

Unions on both sides of the Atlantic had their first significant confrontation with music technology in the late 1920s when sound film arrived. The early years of cinema had been a boom time for musicians as many cinemas used their own pianos, organs, or orchestras to accompany silent moving pictures. Almost all cinema musicians' jobs were at risk in 1927 when sound film arrived. Railing against this "Talkie Menace," one union member wrote:

> The development of anything with scientific or economic value cannot be hindered . . . This includes the development of machines of all kinds, but the machines producing material products necessary for the well-being of man are an entirely different proposition from machines which enter the domain of art, hence an intellectual field, and are intended as a substitute for the personal services of an artist . . . Were the future to hold the possibility of turning our theatres into music mausoleums with soulless photographic reproductions of the services of musicians and actors, the culture of art and music would suffer.[21]

The earliest screenings of talking pictures were far from edifying, as projectionists struggled to synchronize soundtracks on Vitaphone discs with actions on the screen. In 1927, the Musicians' Union confidently predicted audiences would reject such low-quality offerings in favor of live musicians: "It remains to be seen whether, when the novelty wears off, the patrons of the theatres will be satisfied with this dehumanised form of entertainment." Members were forbidden from "taking part in the production of sound films, movietones, vitaphones, panatropes or in the making of records for film accompaniment,"[22] while the union waited for audiences to come to their senses and reject talking pictures on aesthetic grounds. In March 1930, a cartoon appeared on the front of the *Musicians' Union Journal*, titled "Picture of a Cinema Turning to Sanity" (Figure 4.2). It showed an audience clamoring to return to deserted cinema where "canned music" had been thrown in the bin and a "real, flesh and blood orchestra" had been reinstated "due to popular demand." The Musicians' Union distributed leaflets to cinemagoers, explaining how their hard-earned cash, which once paid the wages of local musicians, now lined the coffers of large movie studios, most of whom were in America. One campaigner confidently wrote: "By propaganda, the Americans have forced the Talkies down the throat of the public and by *our* propaganda—which we might call an emetic—we have forced them to throw them back again." Meanwhile, their associates in the American Federation of Musicians took out half-page newspaper advertisements, warning against the "Talkie Robot."[23] An advertisement in the *Chicago Tribune* showed a robot chained to a harp, with Cupid fluttering out of reach.

Despite this fighting talk, by 1930 the union was already out of step with public opinion. As sound film technology matured over the late 1920s, the public appetite for talking pictures grew unabated, even if ticket holders did miss the services of a good live musician. Responding to public pressure, London's Paramount and Gaumont cinemas briefly reinstated their orchestras in 1931 but laid them off again, just a few weeks later, when they were unable to cover their running costs.[24] That same year the Musicians' Union admitted defeat, writing "we must realise these devices have come to stay." Jobs for musicians in cinema had been short-lived—as James Kraft notes: "What technology gave it eventually took away."[25] Kraft wonders if this outcome, which led to mass unemployment among musicians at the time of the Great Depression, made musicians particularly wary of other technological innovations to come.

Considering their failure to kill off talking pictures, it's interesting to see the Musicians' Union deployed a similar line of attack against the drum machine fifty years later. Of course, in both cases musicians were contending with an abrupt change in the means of production of music—one that might result in economic hardship. Sound film takes human musicians out of the supply chain, once a single take has been recorded. It enables perfect replicas of a performance to be heard time and time again, with no recourse to live musicians. Similarly, the early drum machine supplies never-ending perfect copies of a rhythm, hardwired into its electromechanical or electronic box, with no need for the services of a drummer. The automatic replication of musicians' efforts was something the Musicians' Union had always found irksome. In an article titled "Cultural Diminuendo," an anonymous union member in 1959 wrote:

> If musicians were selfish—if they thought only about the interests of their own profession—they would refuse to continue performing for recording, broadcasting and television. They would insist on recreating the conditions that existed only a generation ago, when a musician had to personally be present to give his

Strictly Private. **For the Use of Members Only**

MUSICIANS' UNION MONTHLY REPORT

General and Registered Office: 135, MOSS LANE EAST, MANCHESTER

Telegraphic Address: "AMUSE," Manchester. Tel.: Manchester 5489 Central.
Tram Cars Nos. 10, 11, 12, 22, to Prince of Wales Hotel.

To save disappointment interviews can only be had by appointment.

LONDON DISTRICT BRANCH OFFICES: 17, NEW BURLINGTON ST., REGENT ST., LONDON, W.1
Telegraphic Address: "AMUSOLOGIA PICCY." *Telephone:* 4486 GERRARD.

No. 3. NEW SERIES. MARCH, 1930

THE TURN OF THE TIDE.

FIGURE 4.2
The Musicians' Union predicts the public will soon tire of sound film and reinstate live musicians, March 1930. Courtesy of the Musicians Union Archive, University of Stirling.

> performance—when the performance was inseparable from the performer, was transitory, and left behind no tangible commodity that could be bought and sold.[26]

But of course, as the union themselves had noted, music is "in the domain of art," making it an unusual product—one that is subject to more than raw economic judgments. Perhaps this explains why the union's arguments against new music machines, however rational in economic terms, carried so little weight. Cinema sound was appreciated *because* it was different to a live orchestra—it could give the audience an encounter with an exotic or remote location, for example, or present them with the voices of recording artists they would never have the chance to meet in the flesh. And as we see in the rest of this chapter, drum machines were ultimately embraced because of their unrelenting, "stilted" performance. They offered something distinct from a human drummer. Their sound perfectly fitted the machine aesthetic that was developing in popular music in the postindustrial age—a development that defied the union's predictions.

Convenience and Autonomy

As the Musicians' Union suspected, many early adopters of the drum machine were touring musicians looking for a luggable alternative to a drum kit. The drum machine was one of many electronic instruments that first attracted musicians because it was so convenient. The same can be said of substitutes for other hefty instruments, most notably alternatives to the piano and Hammond organ. Arguably, many substitute machines were poor-quality mimics of larger instruments but were tolerated simply because they were easier to lug around. Over time, players and listeners habituated to these mimics, either because they grew to like their qualities, adapted their music to suit them better, or simply became inured to their deficiencies.

The Farfisa Combo Compact electric organ (1964), for example, was hugely popular, despite its thin timbre, as it was advertised as the organ "for the band who's travelling light." Transistorized circuits, originally developed for electronic accordions, enabled Farfisa to make this instrument small and light. Inside a Hammond organ, notched metal tonewheels spin in front of electromagnetic pickups to create a harmonically rich, reedy sound. The Farfisa, on the other hand, simply filters two superimposed square waves (with a 67 percent pulse width), creating a thin imitation of the true Hammond sound.[27] The Farfisa was a fair approximation of the Hammond at a fraction of the weight and price, and the makers were keen to stress this. One advertisement showed hipsters effortlessly carrying a Farfisa down precipitous steps and into a rowing boat, destined for a gig on a faraway shore. As more bands used the Farfisa as a budget Hammond replacement, its sound became tightly associated with the kind of music they were making—R&B and soul. Arguably, the Farfisa sound (and the sound of its more luxurious counterparts, by companies such as Vox) became associated with 1960s soul music as much as the

Hammond itself. A Farfisa provides the unmistakable organ solo at the beginning of the Percy Sledge hit *When a Man Loves a Woman* (1966), for example.

Of course, keyboard players didn't favor the drum machine just because of its size. The drum machine gives the nondrummer autonomy to create dance music, without fretting over the costs, tastes, and other whims of another musician. In an interview with web magazine *Motherboard TV*, contemporary musician Moby, an avid collector of drum machines, wryly noted:

> Logistically being in a band is quite difficult, because if you have four people in a band in the suburbs . . . one of them's drunk, one of them is at work at the mall, the other's with his girlfriend . . . whereas . . . electronic instruments, they're always waiting, they never complain, they don't have girlfriends . . . [It's] that seduction of having an unlimited sonic palette with . . . musicians who are always there waiting for you. [28]

Musical pioneer Sly Stone may have been looking for this autonomy in 1970 when he worked alone on the album *There's a Riot Goin' On*.[29] At that time, Sly and the Family Stone were heavily under the influence of illegal narcotics, and this was wrecking their ability to work together. Distancing himself from fellow band members, Stone retreated to a friend's mansion in Bel Air and pieced the album together on a TEAC multitrack tape machine. Recording sessions took place in his attic and in a Winnebago parked in the street. The process was chaotic, with Stone layering many parts together to create each song. Frequently, Stone would call in a backing singer or musician, record them on the TEAC, then wipe off their contribution the next day, as he'd changed his mind about a track. Stone overdubbed, erased, and rerecorded to such an extent, he wiped much of the ferric oxide off the tape (allegedly, Stone's tape looked "translucent" by the time he handed it to the mastering engineers.[30]) This may account for some of the album's subtly distorted, "muddy" sound quality, which perfectly fitted the foreboding themes of the tracks. In songs such as "Spaced Cowboy," excessive overdubbing left many vocal parts barely intelligible and instrumental lines hard to distinguish in the mix. Yet the songs were saved from chaos by the drum track, which acted as an audible backbone, enabling Stone and his guest musicians to create a complex yet coherent aggregate performance over many subsequent takes. While some tracks feature the work of drummer Greg Errico (who allegedly quit halfway through the project) or his replacement Gerry Gibson, others were wholly or partly played on the Maestro Rhythm King, a budget drum machine that Stone called his "funk box."

The Rhythm King was a solid-state machine, not much larger than a bread bin and light enough to pick up with one hand. Adjusting the tempo dial varied the resistance (and therefore the period) of the machine's central R-C timing circuit. The machine also had a "balance" dial, which affected the relative volume of the high hat and snare drum, and a single shot button for each drum sound. Unlike the SideMan, which used a rotary selection switch, the Maestro had push buttons—one for each rhythm pattern, and more than one rhythm could be played at a time. Pressing the "bolero" and "go-go" buttons simultaneously, for example, triggered a polyrhythm, comprising these two distinct rhythm patterns playing at the same tempo. Without a machine, polyrhythms like these were beyond all but the most virtuosic drummers.

Talking to journalist Steve Peacock in 1972, British psychedelic musician Arthur Brown summed up what it meant to have a machine with these capabilities:

> There's one thing that limits all rock bands, and that is the fact that a rock band is built around a drum-kit. The drum-kit is limited by what the hands and the feet of the drummer can do. The Rhythm Ace is equivalent to 200 ordinary drum kits. It frees the whole scope of rhythm and lets you get into patterns of rhythm that you just can't get with a drum kit.[31]

Brown surprised Peacock when he introduced his new drummer "Ace Bentley"—a Rhythm Ace machine that had been packaged by the piano company Bentley for sale in the United Kingdom. This drum machine had taken over the reins when Brown's decidedly human former drummer allegedly absconded with the bass player's wife. Brown had made his name with the track "Fire" (1968), and by the early 1970s he was experimenting in the studio, making electronic rock. The drum machine fitted perfectly into his evolving music style—his guitarist Andy Dalby was particularly taken with its ability to play softly:

> We can get a better sound whenever we're on stage too because everything's much quieter and you can actually hear what's going on. You don't have that acoustic problem of having to play as loud as the drums to get over before you start playing the music. It's a bit like being in the studio—you can control it.[32]

Ace Bentley was featured heavily on *Kingdom Come* (1973), a studio album that proved fiendishly difficult to perform live. One problem was the drum machine's lack of *rubato*: "The drum machine had its own discipline. The bass player regularly used to get a headache because its timing was so rigid."

It was the timbre of the drum machine that attracted musician Wolfgang Flür, when he found one languishing in Kling Klang, Kraftwerk's studio in Düsseldorf.[33] Flür had recently joined Kraftwerk and was rehearsing with Florian Schneider and Ralf Hütter for an upcoming TV appearance. It was September 1973 and Flür, a skilled drummer, had only recently joined the band. He was mortified when his new bandmates presented him with their acoustic drum kit, which was little more than a toy. The prospect of a TV appearance with such a low-quality kit made him very intrigued to find a Maestro Rhythm King lying in the corner of the studio—it was something the others had bought for use as an occasional stand-in drummer. Flür liked one or two of the Maestro's rhythms but was more taken with the electronic sounds that made them up, most notably the synthetic bass drum, which he described as "rich and dry," perfect for playing over Hütter's bass horn.

Working with Schneider, Flür hacked into the push-button switches that triggered single-shot sounds and connected them to his own external triggering device. This consisted of nine copper plates (one for each sound), each of which could be activated by striking it with a brass tube, which was connected to a low voltage. The various sounds of the Rhythm King were triggered when he struck the plates with this tube. Thus, Flür could use his conventional, human drumming technique to trigger electronic sounds. He could create an electronic drum part that had the expressive *rubato* of a human player.

Some musicians may only have used the drum machine as a cheap and convenient substitute for a human drummer. And a few drum machine companies attempted to make these machines more lifelike by adding "humanizers"—buttons that gave a little lifelike randomness to the time keeping. But clearly, within fifteen years of the sale of the first Wurlitzer SideMan, adventurous musicians were already viewing it as a desirable addition to their kit, a device that offered something distinct from a human player. Musicians such as Stone, Brown, and Flür relished its inhuman drumming skills or its distinctive, electronic timbre. These artists championed it as a desirable addition to the creative musician's arsenal.

The drum machine may well have put individual drummers out of work. But it's hard to tell if drum machines reduced the overall employment of drummers. The drum machine can't be isolated from other influences, such as changing musical tastes, the growth of the discothèque, and the arrival of machines such as the synthesizer. But we do know that live percussion has not gone entirely out of fashion in the forty years since the first drum machine sale.

The Chamberlin and the Mellotron

From the late 1950s, jukeboxes, then discotheques and radio DJs, helped to move us toward an era of "transmitted music," where recorded music became commonplace and live performance comparatively rare. However, one development eclipsed all others in its potential to strip live performers from the music *creation* process. The device in question was the sampler—a machine which records segments of sound and allows you to manipulate and access them instantly, so they can be replayed and recombined in any order. A sampler can be used to record the notes of a saxophone, for example, and play them back using a keyboard. Given its impact on music making, it's surprising to discover the Musicians' Union expressed no great concern about the sampler when they inspected an early example in the mid-1960s.

First exhibited in London Olympia, 1963, the Mellotron is widely thought of as the first music sampler to go on sale. But the Mellotron is actually a refinement of an earlier instrument, designed and built by Harry Chamberlin, an electronic engineer and saxophonist based in Upland, California. The development of the Mellotron starts with Chamberlin's Model 100 Rhythmate,[34] a drum machine he designed and built around 1947. This played back drum tracks that were prerecorded on loops of tape. The Rhythmate contained fourteen loops of one-fourth-inch tape, arranged side by side. Each of these carried three different drum tracks, providing a total of forty-two different rhythms. As the Rhythmate had a single tape head, it could play only one of these loops at a time. A long slider, on the control panel, moved the tape head from one loop to another, and a smaller lever shifted the tape head between the three tracks (this interface wasn't perfect—users often reported problems with crosstalk). The machine also had a tempo dial that varied the speed of the tape capstans, shifting the tempo and pitch of the rhythms. Chamberlin built only around ten Rhythmate units—all were made in his family-run workshop.

Chamberlin moved closer to building an instrumental sampler in 1949. The idea came to him while he was recording a tune on his electric organ. Some years later, he described this moment of inspiration to *Crawdaddy* magazine:

> I bought myself a tape recorder and laid it on the bench next to me. And I was putting one finger down . . . and I said "For heaven's sake. If I can put my finger down and get a Hammond organ note, why can't I pick a guitar note or trombone note and get that under the keys somehow and be able to play any instrument. So long as I know how to play the keyboard, I can play *any* instrument!"[35]

Inspired, Chamberlin recorded individual notes of a handful of instruments, including the saxophone, spliced them onto separate lengths of tape, and arranged the tapes so each one lay under a single key of the organ. Every tape would be controlled by its own key. This was no easy technical feat as each tape needed its own spooling mechanism and playback head, independently controlled by a key press. But when the machine was running, he could "play" the sampled (recorded) instruments simply by running his hands up and down the keyboard. The temporal envelope of an instrument helps to give its distinctive sound. A flute note, for instance, has a slower attack and can be sustained far longer than a note plucked from a guitar string. With this in mind, Chamberlin knew his machine had to replay each recorded note from the beginning, in order to make a convincing imitation. A continual looping sound wouldn't do. So he devised a rewind mechanism, using rubber bands to spool rapidly to the start of a tape, every time a key was released (Chamberlin's mechanism was a precursor of the one used in the Mellotron, which is detailed on pages 111–113). The first of Chamberlin's eponymous samplers stored six instrumental sounds, arranged longitudinally on tape, and it was exhibited at the National Association of Music Merchants annual fair in 1956. He was deluged with orders at the end of the show.

How the Chamberlin inspired the Mellotron is open to some dispute. But it is widely believed to involve Bill Fransen, who worked as Chamberlin's salesman in the early 1960s.[36] At the time, Chamberlin's company was facing some technical hitches with the hundred or so Model 200 machines they had sold. Instruments played with an uneven tone and volume as the machine used unmatched tape heads. And the mechanics themselves were lousy. Allegedly, the instrument had a 40 percent failure rate, as far too often tapes snarled up as they spooled from one sound bank to another. In 1962, Fransen took a couple of Chamberlins with him to England in search of some technical help. He met the Bradley brothers, Leslie, Norman, and Frank, who ran an engineering company in Birmingham, specializing in electronics for tape recorders and amusement machines. From this point, accounts vary, but Fransen may have made this trip without Chamberlin's knowledge. When Fransen asked Bradmatics for seventy matched tape heads, the brothers wondered if they were sourcing components for some kind of musical instrument, and they were curious enough about Fransen's inquiry to build their own version of the Chamberlin in collaboration with him. This became the Mellotron Mk I (*Mel*ody Elec*tron*ics). Bradmatics were under the impression that copyright for the machine belonged to Fransen, and they were surprised when a furious Harry Chamberlin arrived to set the record straight some months later. In 1966, Chamberlin and the Bradleys' new company Mellotronics resolved their dispute over ownership, when the Bradleys allegedly paid Chamberlin a sum of $30,000. The two companies, Chamberlin and Bradmatics, continued to collaborate, here and there, on master tapes for their two machines. When the Mark I Mellotron was exhibited in 1965 in association with the tape manufacturers Scotch, Bradmatics received a number of inquiries, not all of them musical.[37] The BBC commissioned one to store and playback spot sounds (short sound effects) for programs such as *Z-Cars* and *Doctor Who*. Another inquirer wanted one to play reference recordings of heart murmurs.

Selling the Mellotron

Unlike the drum machine, the Mellotron wasn't designed as a gigging instrument but as the ultimate home keyboard. Aimed squarely at rich people who had more passion for music than talent, the sales brochure around 1965 showed a Mellotron in a large and luxurious living room. This instrument was remarkably expensive—the Mark II was just a little shorter and stouter than an upright piano but sold for £975—roughly six times the cost.[38] The Bradleys manufactured Mellotrons under the company name Streetly Electronics but sold them through Mellotronics, which they set up in partnership with bandleader Eric Robinson and his son-in-law, the magician David Nixon. Robinson explained the virtues of the instrument:

> All over the world there are countless people who have always wanted to play like a professional, yet either there wasn't the time to learn, or else those fingers would never hit the right notes at the right time. And so it happened that in many homes the piano became the stand for photographs, and the family turned to the record player, the radio and the television for their music . . . Now, with the fabulous Mellotron,

> anyone with the slightest ear for music can command his own orchestra—simply by using two fingers and a thumb—producing a wealth of orchestral sound never before obtainable from a single keyboard . . . I regard the Mellotron as the greatest development in home entertainment since television.[39]

Nixon and Robinson demonstrated the Mellotron in a British Pathé newsreel in 1965. Here, Nixon described the Mellotron as a "musical computer" and again, depicted it as the perfect instrument for the willing but unskilled amateur.[40] "I'm a frustrated musician," explained Nixon. "I need this, you see—I've never been able to play the piano." He then demonstrated his "party piece," a quickstep version of the jazz standard "Bye Blue Blues," "with two fingers, and nothing up my sleeve!" Nixon played banjos, organ, trombone, and percussion, all supplied by the Mellotron. At the end of the newsreel, a professional pianist, Geoff Unwin, took over the instrument and played some dance music that really put the Mellotron through its paces—but the overall message was clear: Even if you could only play with a couple of fingers, the Mellotron could make you sound like a maestro—a whole orchestra, in fact.

How the Mellotron Works

On the outside, the Mark II Mellotron looked very much like an upright piano, although it had two thirty-five-note manuals, arranged side by side, in its mahogany veneered box. Around 300 Mark II Mellotrons were sold between 1964 and 1968, and most of the earlier Mark I models were upgraded to Mark II. Aimed squarely at the amateur domestic market, the Mellotron was equipped with sounds that suited popular dance music.[41] These were all prerecorded on magnetic tapes, stored inside the machine. Equipped with valve amplifiers, the Mellotron had a bandwidth of only 8kHz, which gave it a mellow, slightly muffled tone.[42] The instrument also had a built-in spring reverb—an electromechanical device that could be used to add reverb to "lead" instruments, such as flute, clarinet, mandolin, and French accordion.

Lead instruments were played on the right-hand keyboard; "background" instruments, such as cello and violin chords, celeste and trombones, were played on the left. So were rhythms such as bossa nova, quickstep, and foxtrot. The left-hand keyboard could also be used to play a few spot sounds. "You'll find these a great asset when making your family home movies," explained the user manual. Taking a closer look at the Mark II Mellotron, we can see how sounds were activated when you pressed the Mellotron's keys: Inside the Mellotron Mk II are seventy tape machines, side by side. Figure 4.3 shows the workings of one tape machine. It carries ⅜-inch tape, which is wound over a front spindle, then between a pinch roller and capstan, over a rear spindle and into a cassette box. The pinch roller is fixed to the underside of the key, and the tape is arranged so there is a six-foot (1.8 meter) loop of slack. This is wound around an extra set of rollers (the "bottom rollers" in the diagram), which is attached to a spring.

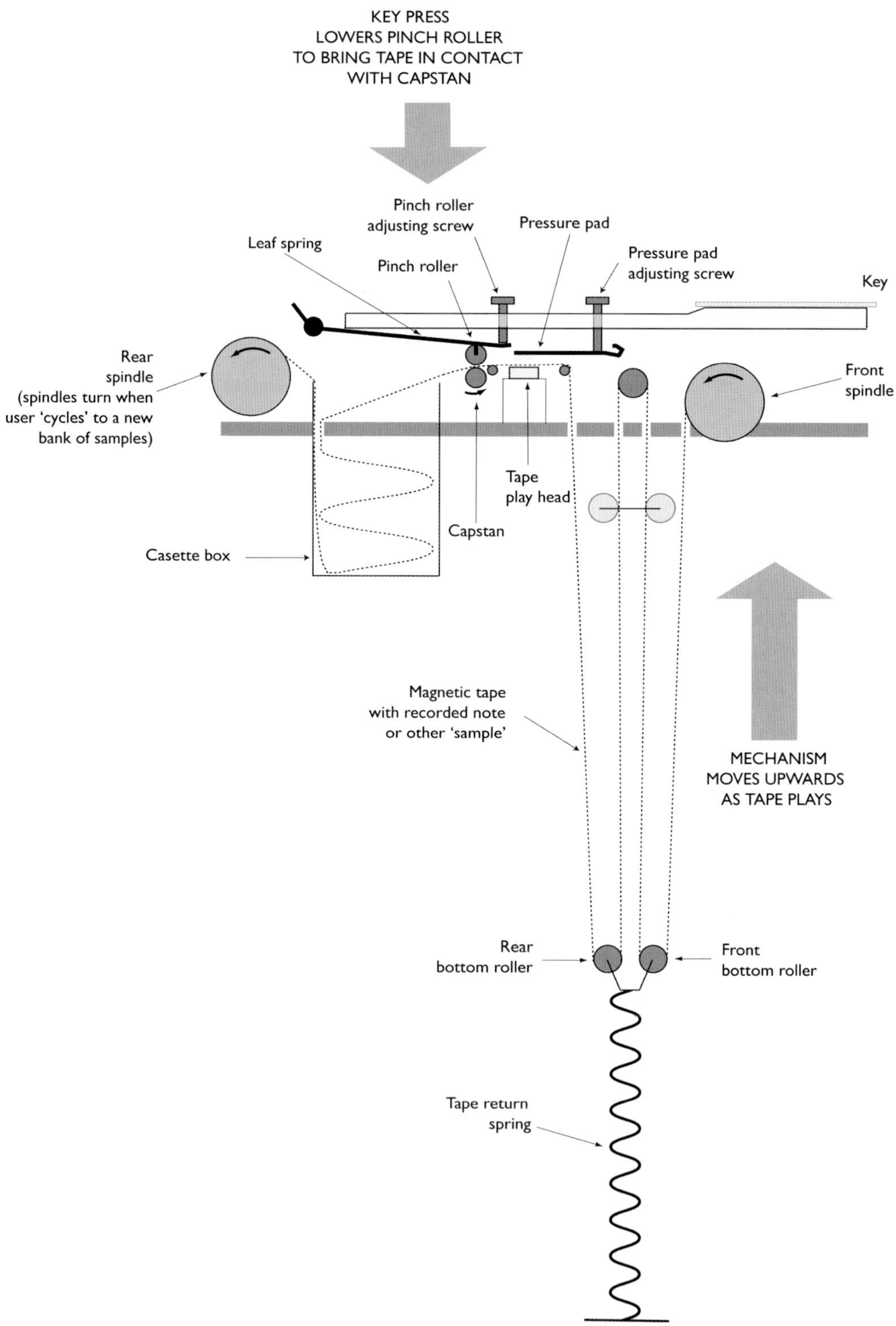

FIGURE 4.3

The tape mechanism coupled to each key of the Mellotron Mk II. Courtesy of Sarah Angliss.

The capstan turns at constant speed (in theory, at least). As soon as a player presses a key, the pinch roller pushes the tape toward the capstan, which feeds the slack tape through the machine and into the cassette box. Here, the tape can gather in loose folds without tangling. At the same time, a pressure pad pushes the tape toward the play head. As soon as the key is released, the pinch roller and pressure pad retract so the tape stops playing. The spring attached to the bottom rollers also retracts at the moment of release, pulling the slack tape out of the cassette box. This action happens in a fraction of a second, so the recording is ready to be played, from the beginning, as soon as the key is pressed again.

As the slack is six feet long and the tape runs at 7½ inches (0.19 meters) per second, the machine can play notes that are up to ten seconds long. Each length of ⅜-inch tape contains three tracks (for instance, marimba, violin, and trumpet). You can switch between these or mix them together using "channel" buttons on the front of the Mellotron. This moves the play head a few millimeters left or right to select different tracks. The tape can also be "cycled" backward or forward to a new six-foot section, where three entirely different sounds are stored. During cycling, the front and rear spindles turn, spooling the tape to the new sound bank. The tape has six different sound banks, each with three tracks, so there are eighteen different sounds on each key of the Mellotron.

Of course, the Mellotron's sound banks can be changed by installing a new set of tapes in the machine. But this is no easy undertaking, as the tape is a nonstandard width (⅜-inch) and every one of the seventy tape machines has to be threaded with its own length of tape. The Bradleys offered a service to trim and install standard one-half-inch tape,[43] but it's unlikely this had many takers. The Mellotron 400, which arrived in 1970, had tapes installed on a cassette that could be removed simply by undoing a few screws (Figure 4.4).[44] This made it easier to swap sound banks, although it would never match the convenience of calling up sound banks on digital samplers, which arrived in the mid-1970s (see later).

The Mellotron Makes Its Mark

The Mellotron might have been a footnote in the history of popular music if it had remained a domestic instrument. But within three years of the first Mellotron sale, it was picked up by musicians including The Moody Blues, Yes, Jimi Hendrix, King Crimson, and The Beatles, all of whom used it on memorable tracks. It was the Beatles' use of the Mellotron on "Strawberry Fields Forever" (released February 1967) that caught the attention of journalists from *Melody Maker*:

> "Listen to those trumpets—fantastic." That's what Beatles fans were saying when they first heard the group's incredible "Strawberry Fields Forever" with its dramatic orchestrated arrangement. Then it was revealed that helping to make all those extra-Beatle noises is a remarkable instrument called The Mellotron. The Mellotron is slowly taking place in the British music scene . . . and the Beatles have been the first to make successful use of it on a hit single.[45]

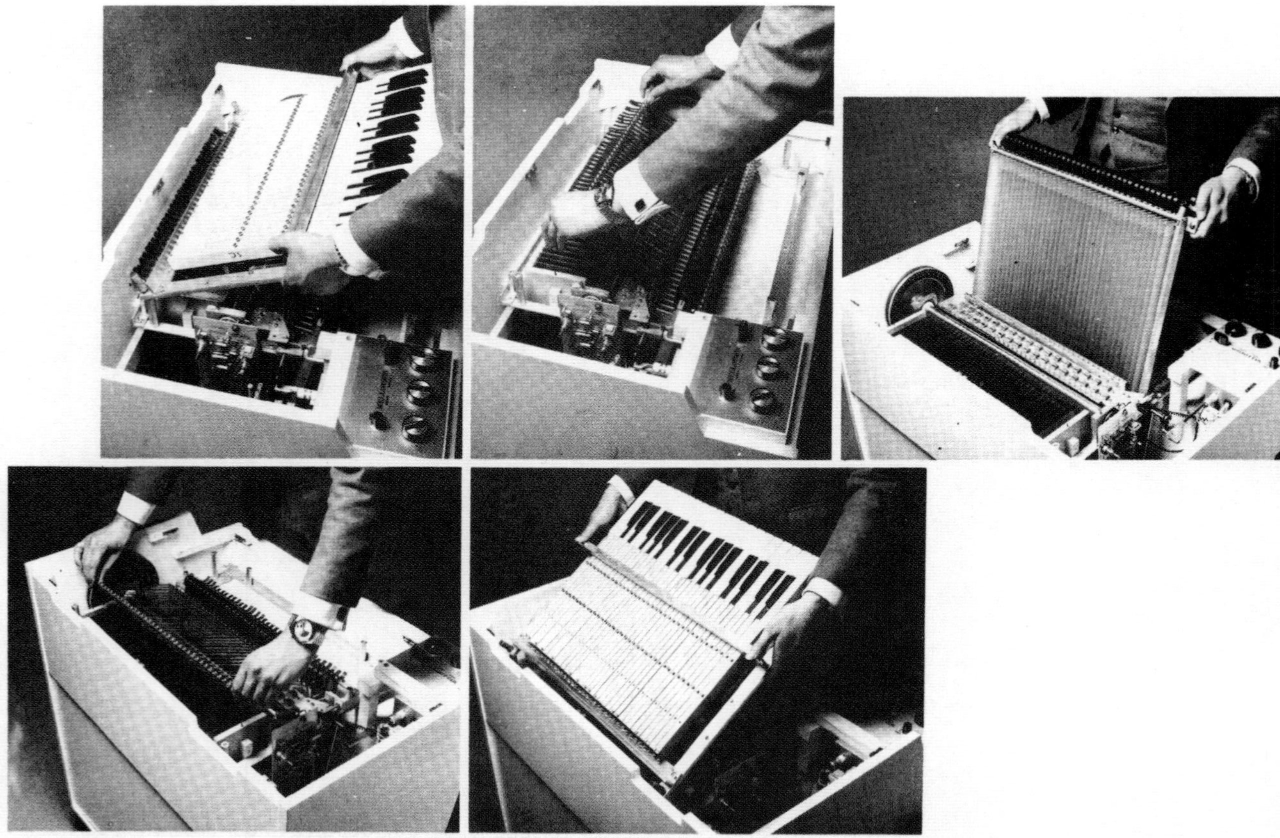

FIGURE 4.4

Tape cassettes in the Mellotron 400 could be swapped in under two minutes, according to the manufacturers. From the Mellotron brochure (reproduction courtesy of Streetly Electronics File T/1992-93). Courtesy of the Science Museum. Images available from Science & Society Picture Library.

The Mellotron had been used to supply brass sounds on this new track, penned by John Lennon, and also provided a memorable introduction. Working with producer George Martin, Paul McCartney created these opening notes when he abstracted some chords from the song and played them using the Mellotron's flute voice.[46] The result had an odd character: Made from disparate, sampled notes, it lacked the *legato* of a real flute but it did, nonetheless, sound unmistakably flutelike. The Mellotron evoked the flute but had a sound all of its own.

Nine months later, a Mellotron provided lyrical, orchestral string lines in the Moody Blues' hit "Nights in White Satin" (November 1967). Their keyboard player, Mike Pinder, became a Mellotron devotee after working as a quality inspector and tester at Streetly Electronics. When he left the company to concentrate on his band, the Bradleys helped him source a Mellotron Mk II that was standing, unused, in the social club of the Dunlop Tyre Factory, Birmingham.[47] Spearheaded by bands like the Moody Blues, musicians with plenty of money (and bravado) attempted to take Mellotrons on the road. But this domestic instrument was ill suited for touring. Described as "the roadies' nightmare," this hefty machine invariably needed fixing as soon as it came out of the van. The azimuths of its seventy tape heads often had to be realigned, individually, after the instrument had been moved. After that, its performance could still be idiosyncratic.[48] Notes would stop working halfway through a gig, as one of the seventy delicate tape mechanisms failed, and the capstan labored if too many notes were played simultaneously, making the pitch slip downward. Heat, humidity, and stage smoke were also known to make the pitch fluctuate. Speaking on the BBC documentary *Synth Britannia*, King Crimson keyboardist Ian McDonald said he performed their album *In the Court of the Crimson King* (1969) on the Mellotron at fifty or sixty gigs and only heard it in tune once.[49]

The Mellotron was clearly pushing tape technology to the edge of its capabilities. But bands seemed willing to overlook the resulting operational headaches as they were so taken by its ability to play back recorded (sampled) sounds. Pinder always freighted a back-up Mellotron from gig to gig, just in case he hit any technical problems.[50] He was also one of just a few musicians who went to the trouble of making their own samples for the Mellotron—he used it, for instance, to replay sounds of a Moog synthesizer. McCartney was also known to make up his own samples,[51] for instance, to play the bagpipe sounds in his solo record "Mull of Kintyre" (1977). When the Mellotron 400 arrived, some players carried several cassettes with them, and fearless roadies would swap between cassettes halfway through a set.

The Mellotron Myth

The Mellotron seems to have a special place in musicians' psyche, particularly in the United Kingdom—perhaps because it has been used on so many unforgettable British music tracks. It's remembered fondly as the "primordial sampler" that left many roadies and keyboard players sweating. But it's also often cited as the instrument that was "banned by the Musicians' Union." According to popular myth, the union banned the Mellotron from BBC recordings because it was putting jobs at risk. This interpretation of the Mellotron's history is summed up very well on the sleeve notes of the Mellotron compilation album *The Rime of the Ancient Sampler* (1993):

> At first the Musicians' Union kicked up a real fuss banning the [Mellotron's] use. Mike Pinder, however, somehow managed to feature two of the instruments on a TV special of the Moody Blues, but otherwise a lot of wrangling went on, and the MU was defied by many musicians who were hooked on Mellotrons.[52]

Considering their response to sound film and the drum machine, we would expect the Musicians' Union to be highly critical of the Mellotron, an instrument that could replace not just a drummer but almost any instrumentalist. So it's surprising to discover that the union didn't ban the instrument at all. There was a minor dispute in 1973 involving the Mellotron at the BBC Lime Grove Studios. But Stanley Hibbert, sessions organizer for the Musicians' Union at the time, described it as a "storm in a teacup" that was resolved amicably, simply by moving an offending Mellotron from one studio to another.[53] Hibbert wondered if the storm was generated by people who had taken the claims of Mellotronics a little too seriously.[54] As early as November 1966, a delegation from the Musicians' Union visited the Bradleys to check out the Mellotron and gave it a lukewarm review:

> The Mellotron is new, and like many new developments has its weaknesses. A five-note chord played high in the treble with the string tone indicated hardly sounded like a string section, and a "fanfare of brass" lacked the precision-like sound of live performers.[55]

Unlike Nixon, they thought the instrument was of little use, unless it was in the hands of a skilled keyboard player. They concluded: "We believe that the threat to the livelihood of musicians by the 'Mellotron' is no greater than that by any electronic organ."[56] So the Musicians' Union enabled the Mellotron to usher in the sampling era with no great protest. Confusingly, in the mid-1970s, the Musicians Union encapsulated some more general thoughts on imitative instruments by drafting this directive:

> Instruments and devices incorporating pre-recorded sounds or producing sounds by electronic means must not be used to replace or reduce the employment of conventional instrumentalists in circumstances where these may be reasonably expected to be used; it is understood, however, that these devices may be used to produce sounds that cannot be produced by conventional instruments.[57]

According to this logic, a Mellotron should be out of bounds. But the instrument slipped under the radar, perhaps because of its domestic roots, perhaps because it seemed too expensive and unwieldy to offer "a little bit extra, at no extra cost."[58] Reading their paean to a bygone age when "the performance was inseparable from the performer,"[59] it is odd that the Musicians' Union were untroubled by an instrument that could easily do the work of an entire string quartet with one pair of hands. The Mellotron—and the many samplers that followed it—presented a textbook example of mechanization. Here was a machine that could be bought by cash-rich studios and used to replace session musicians.

In the 1970s, Barry Frederick, a New York session musician, offered himself or his Chamberlin for hire. Speaking to *Crawdaddy* magazine, he was well aware of the economics:

> If I'm not playing on the session, they pay $150 for 24 hours rental. In 24 hours, you figure how many musicians you'd have to pay. Union scale is at least $100 for three hours for each guy. So you'd have one guy playing 'cello, one playing violin and one flute, that's at least . . . $2,400 for 24 hours.[60]

Of course the Mellotron affected music quality when it replaced session musicians in the studio—but its impact wasn't entirely negative. In offering a tireless substitute for a session musician, the Mellotron made it practical for musicians to experiment for long hours in the studio and develop new kinds of music. They could work solo, developing complex harmonic or melodic ideas, without needing to communicate their ideas to other players. Perhaps it was the desire to re-create these complex studio pieces live that prompted so many bands to take their Mellotrons on tour. Once again, the Musicians' Union were taken by surprise when musicians were so excited by the Mellotron's musical capabilities, they were prepared to haul this behemoth from venue to venue, overlooking its tuning problems and other weaknesses as they nursed it through gigs.

The Growth of Sampling Culture

The Chamberlin and Mellotron gave musicians their first taste of sampling, a compositional technique that would spread through popular music culture from the 1980s onward. The two wellsprings of today's sampled music culture are the first digital samplers, such as the Fairlight (1975) and Synclavier (1978), and the turntable wizards of the block party scene that emerged in South Bronx, New York, around the same time. In many respects, these two sources had very little in common. Early digital sampling was a phenomenally expensive venture, championed by the richest artists and studios, using samples stored on digital memory. Vinyl sampling was a low-cost, grassroots enterprise, remixing music recorded on analog gramophone records. It's hard to claim a direct lineage between either of these sampling methods and their tape-based ancestor, the Mellotron. Digital and vinyl sampling are fascinating, though, because they presented new challenges to prevailing notions of music—views that were often espoused by the Musicians' Union.

Digital Samplers

Digital sampling arrived surprisingly early. By 1969, the London-based Electronic Music Studios (EMS) offered digital sampling and playback using their MUSYS program, which controlled the whole studio via two PDP-8 microcomputers.[61] The studio was designed by EMS founder Peter Zinovieff and his engineers Peter Grogno and David Cockerell. Using this system, up to 12kB of audio could be stored in RAM—just a few seconds, at a barely tolerable bit depth and sampling rate. It had a hard drive that was 32kB in size (Figure 4.5). PDP-8 computers were also used in the Melodian, a commercial sampler that appeared just a couple of years later.[62] Designed by Computer Music in New Jersey, this monophonic sampler was used extensively by Stevie Wonder in his album *Journey Through "The Secret Life of Plants"* (1979).

The first polyphonic, digital sampler was the Fairlight Computer Music Instrument (CMI), launched in 1975.[63] It was designed by Kim Ryrie and Peter Vogel, Australian inventors who had previously made video-effects machines. They originally intended to synthesize convincing instrumental sounds using digital waveshaping algorithms. But after some experimentation, they decided sampling would yield better results.[64]

FIGURE 4.5
Electronic Music Studios, controlled by the MUSYS system, from a contemporary advertisement (reproduction courtesy of Peter Zinovieff). Courtesy of the Science Museum. Images available from Science & Society Picture Library.

According to legend, the first sound to be sampled and played back on the Fairlight was the bark of an employee's dog.[65] With an architecture based on a Motorola 6800-series microprocessor, the Fairlight Mk I recorded sounds at 8-bits and a maximum sample rate of 24kHz—noticeably less than CD quality (16-bit, 44.1kHz). This specification improved in later models. Sounds were played on a keyboard and were accessed and manipulated using the Fairlight's graphical user interface, displayed on its built-in CRT monitor. Unusually, this came with a light-pen interface, rather than a computer mouse.

Every Fairlight was packaged with dozens of prerecorded samples, stored on a library of twenty-five 8-inch floppy disks.[66] This wasn't a cheap instrument. Prices ranged from $25,000 to $100,000 over the production life of the instrument. Herbie Hancock was an early adopter, as was Kate Bush, who used it on *Never for Ever* (1980), the first commercial album to use the instrument. As soon as they had access to a Fairlight, many musicians were inspired to make their own digital samples. Musicians wealthy enough to get their hands on the instrument experimented with everything from vocal noises to the sounds of pebbles falling down drainpipes—digital sampling freed them to experiment in ways that were unheard of before.

Just like the Mellotron before it, the Fairlight wasn't a perfect mimic of acoustic instruments but presented musicians with its own, distinctive sound. And many of its samples, such as its "breathy choir" ARR1, became tightly associated with early 1980s pop. Perhaps its most famous sample was ORCH5, a single, accented, and staccato chord, played by a full orchestra.[67] ORCH5 was actually a pitch-shifted extract from a recording of Stravinsky's *Firebird*. Musicologist Robert Fink has traced the history of ORCH5 and identified it as the opening chord of the *Infernal Dance of King Kastchei*. Decontextualized and replayed on the Fairlight, this chord made an appearance on Kate Bush's album *The Dreaming* (1982) and played syncopated strikes on The Art of Noise track "Close (to the Edit)" (1984). Appropriately, in the promo video for "Close (to the Edit)," we see people gleefully set upon a violin, grand piano, double-bass, and saxophone with chain saws and angle grinders.[68]

When keyboard player John Robie punched an eight-note chord on the Fairlight, using the ORCH5 sample, this fragment of Stravinsky was put at the center of the emerging hip-hop scene.[69] In 1982, Robie was playing keyboards for Afrika Bambaataa & The Soulsonic Force. Bambaataa, a DJ from the South Bronx, had negotiated the overnight use of a studio—and its Fairlight—and was using it to record some new tracks. One of these was "Planet Rock," a sparse, electronic dance track, mixing ORCH5 with two Kraftwerk quotations: a melody from "Trans Europe Express" (1977) and a facsimile of the drum part from "Numbers" (1981), programmed by Robie on a Roland TR-808 drum machine.[70] An overnight success, "Planet Rock" is widely regarded as the record that kick-started the commercial hip-hop scene. Its mixing of the emcee's vocals with European synthesizer sounds, fragmented digital samples, and a prominent drum machine track reflected Bambaataa's interests (he was an avid collector of unusual sounds). It also captured something of his live performance on the turntables.

Turntablism to Detroit Techno

The emergence of sampled music owes as much to the South Bronx block parties at the end of the 1970s as it does to the early digital sampler. The "blocks" in question were estates of dilapidated apartments, populated by some of the poorest residents of New York, including many black and Hispanic families.[71] Ethnomusicologist Cheryl Keyes has traced the history of these parties.[72] According to Keyes, many young people living in the blocks eschewed the disco scene, which was afflicted by local gang violence. Instead, in the summer, they congregated in local schoolyards and parks, which they used as their space for music and dancing, often hooking their turntables up to the power supply in the nearest lamppost. Activities moved to school halls and community centers in the winter. At these "block parties," disc jockeys, such as Grandmaster Flash, Kool DJ Herc, and Afrika Bambaataa juggled vinyl records on the fly using two turntables to create a collage of music, from fragments of records. One very well-known trick was to find an excerpt of a record with a really good percussion solo. Kool DJ Herc pioneered the live looping of these "break beats."[73] He would play two copies of the record on a pair of turntables. With expert timing, he would reset the needle to the beginning of the break on one record, while he played the break on the other. This way, he could get the break to loop continuously—making the perfect backing for an emcee's patter (which eventually became known as rap).

Other techniques of these turntable virtuosi were "backspinning" (forcing records to play backward toward a musical target), "scratching" (rapidly moving a record back and forth while the needle stays in the groove),[74] and "beat juggling" (rapidly mixing two disparate musical passages from records on different turntables). With these techniques, DJs transformed the turntable from a simple playback device into a musical instrument, one where they could impress the audience with their virtuosity. They were sampling on the fly in the sense that they were accessing recordings, fragmenting them, decontextualizing, and reusing them, and they were reappropriating the sounds they were mixing, along with the gramophone itself. The best DJs turned their art into something of a non-contact sport as they "battled" with each other. As Mark Katz explains,[75] in these battles, DJs fragmented and mixed words from two different records to boast about their prowess and "dis" (denigrate) their musical rivals in "scratch sentences" (virtuosic wordplay like this can still be seen and heard in DJ "battles" today). Bambaataa was known for his skills in bringing unlikely sounds together. His set sampled everything from the *Pink Panther* theme and Sly and the Family Stone to James Brown, The Rolling Stones, and advertisements for the fizzy drink Mountain Dew.[76]

The turntable wizards had a great affinity for synthesizer-based musicians such as Kraftwerk, David Bowie, Gary Numan, and Yellow Magic Orchestra, many of whom toyed with notions of roboticism and alienation in both their music and their stage presence. In particular, songs such as Kraftwerk's "Autobahn" (1974) and Numan's "Are Friends Electric?" (1979) expressed this alienation through sparse, metronomic sounds that made them ripe for mixing with other tracks and reworking on the turntables.

Thematically, these works were also well suited to the turntablists. Hip-hop critic Tricia Rose, reported in the work of Ken McLeod, notes that Bambaataa, like Kraftwerk, Bowie, and Numan, was playing with an alien identity of sorts—his was a flamboyant creation that's been identified as Afrofuturism.[77] Like Sun Ra, Bambaataa used his music, and the sounds within it, to evoke his own extraterrestrial mythology. According to cultural critic Marc Déry, Bambaataa, Sun Ra, and other Afrofuturists were expressing their identity as alien abductees, people descended from robots (forced laborers.[78]) In McLeod's words: "Hip-hop can thus be interpreted as a social emancipation of the robot slave." If early hip-hop does have this agenda, then it's easy to see why its exponents would be attracted to the regimented, "dehumanized" sounds of the drum machine and sampler. Kraftwerk's music, for example, with its fixation on mechanization and its artful imitation of robots, was perfect raw material for the Afrofuturists to sample.

Turntable virtuosi Juan Atkins and Rick Davis (working under the moniker 3070) also took European synth pop and sampled it and remixed it for the Detroit crowd. They called their music Cybotron.[79] Detroit, home of the U.S. motor industry, was also the world center of automation.[80] When it suffered a deep recession in the oil shock at the end of the 1970s, Cybotron offered them stripped-down, dark songs such as "Cosmic Cars" (1982) and "Clear" (1982) that evoked the soundscape of the city and its production lines. The drum machine fitted their aesthetic perfectly, and arguably, it was the re-export of these fragmented and remixed European sounds, as "electro" and later "Detroit techno," that kick-started the techno scene in Europe.

By the early 1980s, hip-hop artists who had learned their craft on turntables got their hands on much more affordable samplers, such as the Akai MPC60 (Figure 4.6). This was based on an earlier device, built by Roger Linn, which was used to sequence digitally sampled drumbeats.[81] Using digital samplers and computers, it was possible to create the sounds of hip-hop and techno, without the skills of a turntable virtuoso. Anyone could pinpoint break beats and other musical fragments, record them, manipulate them, and sequence them to make longer tracks or recall them at the touch of a button. The result was an explosion in sampled music culture and songs based on "loops." Sample culture spread further in the 1990s when the ordinary desktop computer became powerful enough to handle digital audio processing.

In his audio documentary *The Amen Break* (appropriately recorded on a dub plate in 2004), Nate Harrison shows how sampling culture has reworked a single, acoustic performance.[82] As Harrison explains, the Amen Break is a six-second break beat that occurs halfway through the soul record "Amen, Brother," recorded by the Winstons in 1969. "Amen, Brother" was the B-side of their hit single "Color Him Father." This single break beat has been used on dozens of records in hip-hop, drum and bass, raga and techno, from N.W.A.'s "Straight Outta Compton" (1988) to Squarepusher's "Vic Acid" (1997). Invariably, as musicians appropriated this fragment of live performance, they repitched it, looped it, or changed its tempo. Jungle artists sliced it into individual beats that they reordered and raked up in tempo, creating wild rhythmic patterns that bear only a

FIGURE 4.6
An Akai MPC60 sampler and midi sequencer, ca. 1988. Courtesy of the Norwegian Museum of Science and Technology.

passing resemblance to the original. With sampling, musicians were creating works that were decoupled from the live musicians who performed its constituent sounds. But this hadn't led to the "soulless" music that the Musicians' Union had feared[83] but to a new form of human expression. In the words of Katz: "Using turntables, mixers, and lighting-fast hands, turntablists reorganise and recontextualise fragments of recorded sound and, in a kind of musical husbandry, breed rich new meaning from their juxtaposition."[84] Before sampling, musical notes had been the DNA of Western popular music—its units of composition, mutation, and recombination. Now the units of replication could also be recorded sounds, encoded on vinyl discs or as a digital data stream. Note-based composers and improvisers had always quoted one another or based their works around formulaic structures (such as twelve-bar blues), as they wanted their music to be enjoyed by an audience who were conversant with certain musical traditions and who appreciated a nod to other masters. In the high Baroque era, for example, it was fashionable for revered composers such as Handel and Vivaldi to quote each other verbatim in their works. Musicians steeped in sampling culture were arguably engaged in similar activities, using their own forms of quotation: abstracting and reworking excerpts from R&B, soul, and funk masters, such as James Brown and George Clinton, and European electronic pioneers.[85] They too were creating music that could be appreciated by an audience who had grown up listening to venerable artists—and arguably, sampling of earlier works gave their music a strong family resemblance to existing, respected musical genres.

Regardless of its artistic strengths and weaknesses, sampling plunged musicians, engineers, record companies, and the rest of the music industry into uncharted legal and ethical territory. As sampling culture grew from a grassroots activity to a potentially lucrative global phenomenon, many questioned the ethics and legality of sampling. The artists who saw sampling as a creative endeavor found themselves in opposition to those who regarded sampling as theft. Some recording artists were happy to be sampled, perhaps because they appreciated the finished product or because it introduced new audiences to their work. Others were aggrieved to find work had been sampled to create works they disliked or used without their permission. Studio engineers found themselves at the epicenter of the debate. Rumors abounded of session musicians having their performance secretly sampled by unscrupulous engineers who would ask them to run up and down their instrument to "test the mics." Meanwhile, engineers would frequently be called upon to make ad hoc decisions about payments for samples that were made in their workspace. In a U.S. test case in 1991, percussionist David Earle Johnson successfully sued composer Jan Hammer for royalties, when he used a sample of his drumming in the theme tune to the TV show *Miami Vice*.[86] He had no idea the sounds would be used outside the studio. The 1981 film *Diva* (Beineix, 1981) captured prevailing anxieties when it depicted an opera singer who refused to make a single recording of her voice. By the mid-1980s, some musicians were jokingly referring to sample-based music as "high court rock."

Realizing how much money could be made from sampled tracks, record companies stepped up their efforts to assert the copyright of their back catalogue. Tensions arose when musicians were exploiting sampling faster than executives could come to an agreement about the legality. When the British group M/A/R/R/S released their dance-floor hit "Pump Up the Volume" in 1987, Stock Aitken Waterman issued an injunction to stop its sale, arguing it had breached copyright law by sampling their hit "Roadblock" (a song which itself sampled soul legend James Brown). When Waterman wrote an open letter to the press, describing M/A/R/R/S' actions as "wholesale theft," one journalist noted that Waterman's own song "Never Gonna Give You Up" (1987) had itself lifted the bass line from Colonel Abrams's hit "Trapped" (1987).[87]

Those in favor of sampling argued that copyright laws, penned in the eighteenth century, were ill suited to regulate music in the era of digital recording. Organizations such as Copyleft and Creative Commons appeared in the early twenty-first century, offering new ways for people to manage the rights to their works. The Creative Commons formula Sampling 1.0, for example, allowed musicians to give upfront permission for their works to be sampled by others. This progressive attitude to music ownership is analogous to the open source software movement.[88] Interestingly, The Winstons, creators of the original Amen Break, have taken a casual, open source approach to their work all along, never seeking money from the artists who reused it. However, as Harrison wryly notes,[89] at least one commercial sampling company charges people to use their famous drum solo.

Conclusion

Fifty years after musicians tried the first drum machines and samplers, these machines have become an indispensible part of the musician's studio and live performance kit. They weren't rejected as soulless mimics of real musicians but were cherished because they offered something familiar but new. Oddly enough, it's the imperfections of these mimics that made them so attractive to many players. Musicians enjoyed the mellow quality of the valve-amplified Wurlitzer or the otherworldliness of flute chords played on the Mellotron without a trace of *legato*. Even technical quirks, such as the wow and flutter of the Mellotron, as its capstan struggled to regulate speed, helped to engender lifeless, sampled notes with a certain aleatoric charm. Fifty years later, in the era of high-resolution hard disk recording, music connoisseurs will pay good money for "legacy samplers," from software companies such as Propellerhead and Forgotten Keys. These aim to re-create the first drum machines and samplers, with all their audible defects, in software form.

While society continues to argue over the ethics of sampling, it seems this art form is here to stay. As early as the 1920s, the Musicians' Union worried that musicians would lose out as cinemas, radios, and homes moved from live to transmitted music. And when sampling arrived, many expressed concerns about musicians losing ownership of their creative works. Sample-based music may find a way to flourish under toughening regulation. Or perhaps the music industry will adopt a new paradigm, where music has a complex provenance of originators and remixers, rather than a single owner.

Arguably, drum machines and samplers have helped to shift our views of authenticity. Until fairly recently (arguably the mid-1990s), "authentic" music required audible evidence of human agency (or visible evidence). When we compare music making to other handcrafts, such as sewing or pottery, which have been transformed by mass production, we can understand how this assumption arose. Just as a hand-sewn quilt may have fine variations in stitch length and embellishments that could not be executed by a machine, an "authentic" music performance needed traces of *rubato* and other other dynamic variations that are beyond the capabilities of any music machine. These variations were thought to communicate the emotions of the performer, increasing the music's affective qualities. The Musicians' Union argued that drum machines (and to a lesser extent samplers) were deficient because they lacked this variation. But this chapter has shown how people developed a great affection for the inexpressive delivery of these machines, as they suited the mechanized music they were making. For a generation of musicians who have always used drum machines and samplers, the privileging of live human performance may seem unfair. Tom Rowlands from the Chemical Brothers remembers debating this point in 1995, just before a tour:

> A few over-excited record company types thought that we should get a live drummer for live gigs etc. We patiently explained that we + the machine were the drummer—why would we take away the thing that makes the drums interesting. We thought it's 1995 (!)—people would be over the fact they needed to see the physical act of striking a drum when they heard one . . . We don't need to see it.[90]

According to Rowland, the sound they were after was

> a combination of the synthetic with the acoustic, hopefully creating a sort of hyper power drum sound unobtainable by mere man alone!!! We try and make our programmed drums have the loose feel of live playing almost as if the machine is sweating!![91]

Rather than rejecting machines as soulless mimics, if anything, we now have a dash of envy for their capabilities—and perhaps we want to mold ourselves to be a little more like them. As the drum machine proliferated, so did the art of "beatboxing"—using the voice to fake the rhythm patterns and electronic sounds of prized machines such as a Roland TR-808.[92] In software such as Ableton, the rhythmic inflections of soul musicians are ironed out so they can be beat-matched to work in tempo with perfectly timed drum loops. And pitch correction software such as Antares Auto-tune is used to remove the tuning irregularities of a live vocal performance. Whether these developments create exquisite machine music or something soulless is down to personal taste. In the end, the machines are only tools—and the music they make is still only as good as the creative people using them. Speaking in his studio in 1984, jazz musician Herbie Hancock observed: "People blame machines very often . . . 'oh-it's the machine's fault!' How can it be the machine's fault? The machine doesn't do anything but sit there until we plug it in."[93]

Notes

1. *The Mellatron (sic)*, British Pathé, 1965. Newsreel. http://www.britishpathe.com/record.php?id=1962 (accessed August 1, 2010).
2. "The New Wurlitzer SideMan," *Billboard*, May 1960. Advertisement.
3. History of the Musicians' Union. http://www.musiciansunion.org.uk/about-us/.
4. "Don't Fall for It," *Musicians' Union Journal*, 1967.
5. Musicians' Union Monthly Report, March 1930.
6. "The Mellotron," *Musicians' Union Journal*, November 1966.
7. Tom Rowlands, e-mail to author, July 2010.
8. The Wurlitzer SideMan (inv. 1992-101), Mellotron Mk II (inv. 1992-93), Mellotron Model 300 (inv. 1992-97), Mellotron Sound-Effects Console (inv. 1993-735), and Fairlight CMI (1993-654) can all be found in the collections of the Science Museum, London.

9. "Electronic 'Side Man.'" *Popular Mechanics*, November 1960.

10. "The New Wurlitzer SideMan," *Billboard,* May 1960. Advertisement.

11. Mark Brend, *Strange Sounds: Offbeat Instruments and Sonic Experiments in Pop* (San Francisco: Backbeat Books, 2005).

12. "1760–1920: The Rise of the Factory." http://www.spinningtheweb.org.uk/overview/ (accessed November 1, 2010).

13. Edward Baines, reported in "1760–1920: The Rise of the Factory."

14. "Don't Fall for It."

15. History of the Musicians' Union.

16. See also Kolkowski and Rabinovici, this volume.

17. James P. Kraft, *Stage to Studio, Musicians and the Sound Revolution 1890–1950* (Baltimore, MD: The John Hopkins University Press, 1996).

18. Musicians' Union, "Music While You Work," Exhibition text, 1993.

19. Kraft, *Stage to Studio*.

20. Trevor J. Pinch and Karin Bijsterveld, "'Should One Applaud?' Breaches and Boundaries in the Reception of New Technology in Music," *Technology and Culture* 44(3) (2003): 536–59.

21. *Musicians' Union Monthly Report*, March 1930.

22. "M.P.P.A. Important Notice. Disputes Pending with Recording Co's and Talking Pictures," *Musicians' Union Monthly Report*, March 1929.

23. "We Lead—Others Follow," *Musicians' Union Monthly Report*, December 1929.

24. Musicians' Union, "Music While You Work," Exhibition text, 1993.

25. Ibid.

26. Anonymous, "Cultural Diminuendo," *Musicians' Union Monthly Report*, 1959.

27. "Farfisa," http://www.combo-organ.com/Farfisa/index.htm (accessed August 1, 2010).

28. *Electric Independence: Moby*, Motherboard.TV, 2009. http://www.motherboard.tv/2009/10/19/electric-independence-moby (accessed August 1, 2010).

29. "There's a Riot Goin' On," *Super Seventies Rocksite.* http://www.superseventies.com/spslystone.html (accessed August 1, 2010).

30. Various forum members, *Discussion on Gearslutz Forum*, May 2010. http://www.gearslutz.com.

31. Various. *Kingdom Come (Arthur Brown)*, Sleevenotes to CD edition, 2003.

32. Various. *Kingdom Come (Arthur Brown)*.

33. Wolfgang Flür, *I Was a Robot* (London: Sanctuary Publishing Ltd., 2000).

34. David Kean, *A Mellotron History.* http://www.mellotron.com/history.htm (accessed August 1, 2010).

35. Len Epand, "A Phantom Orchestra at Your Fingertips," *Crawdaddy*, April 1975.

36. Kean, *A Mellotron History*; Gordon Reid, "The Rebirth of Cool," *Sound on Sound Magazine*, August 2002.

37. "Scotch Show Mellotron," *Tape Recording Magazine*, May 1965.

38. Kemble & Co. Ltd., *Retail Price List*, 1959.

39. Eric Robinson, *Introducing the Mellotron*, Sales brochure, 1965.

40. *The Mellatron (sic).* British Pathé, 1965.

41. "The Mellotron. Instruction Manual," London, Chappell and Co., 1963.

42. Derek Cook, *How the Mellotron Works.* http://www.echoes-music.co.uk/Mellotron/Start_Here.html (accessed August 1, 2010).

43. Chris Martin, *Streetly Electronics: The Original Mellotron Makers*, http://www.vemia.co.uk/mellotron/ (accessed August 1, 2010).

44. Martin, *Streetly Electronics.*

45. "Paul McCartney Plays Mellotron," *Melody Maker*, January 22, 1967.

46. Danny Danzinger, "Something to Get Hung About," *The Independent*, November 14, 1994.

47. Martin, *Streetly Electronics.*

48. *Synth Britannia.* Directed by Ben Whalley. UK. BBC, 2009. Documentary.

49. *Synth Britannia.*

50. "Instruments Exist Side by Side," *Billboard*, November 6, 1971.

51. Martin, *Streetly Electronics.*

52. Various artists. *The Rime of the Ancient Sampler, The Mellotron Album*, Voiceprint, 1993, CD.

53. Stanley Hibbert, telephone interview, 1993.

54. Hibbert.

55. "The Mellotron," *Musicians' Union Journal*, November 1966.

56. "The Mellotron."

57. Musicians' Union draft directive, ca. 1978.

58. "Don't Fall for It," *Musicians' Union Journal*, 1967.

59. Anonymous, "Cultural Diminuendo," *Musicians' Union Monthly Report*, 1959.

60. Len Epand, "A Phantom Orchestra at Your Fingertips," *Crawdaddy*, April 1975.

61. Graham Hinton, *EMS Electronic Music Studios*, 2002. http://www.ems-synthi.demon.co.uk/ (accessed August 1, 2010).

62. "Questions," *Contemporary Keyboard* Volume 6, GPI Publications, 1980.

63. Greg Holmes, *The Fairlight CMI.* http://www.ghservices.com/gregh/fairligh/ (accessed August 1, 2010).

64. Holmes, *The Fairlight CMI.*

65. Ibid.

66. "The Fairlight Computer Musical Instrument, Users Manual," October 1983. Accessed in stores of Science Museum, London (inventory number 1993-654 Pt 11).

67. Robert Fink, "The Story of ORCH5, or, the Classical Ghost in the Hip-Hop Machine," *Popular Music* 24(3) (2005): 339–56.

68. "Close (To the Edit)," directed by Zbigniew Rybczynski, May 1984. Music video. http://www.youtube.com/watch?v=-sFK0-lcjGU (accessed July 2010). Reported in Fink.

69. Fink, "The Story of ORCH5."

70. A. Baker, "Interview with Bill Brewster and Frank Broughton," January 25, 1999. Reported in Fink.

71. Cheryl L. Keyes, "At the Crossroads: Rap Music and its African Nexus," *Society for Ethnomusicology* 40(2) (1996): 223–48.

72. Keyes, "At the Crossroads: Rap Music and its African Nexus."

73. Keyes, "At the Crossroads: Rap Music and its African Nexus."

74. Keyes, "At the Crossroads: Rap Music and its African Nexus."

75. Mark Katz, *Capturing Sound: How Technology Has Changed Music* (Berkeley: University of California Press, 2010).

76. *History of the Birth of Hip-Hop Culture until 1991, Part 1*, Rapido Television, ca. 1991. Television documentary. http://www.youtube.com/watch?v=k8IO2KHjXZg (accessed August 1, 2010).

77. Ken McLeod, "Space Oddities: Aliens, Futurism and Meaning in Popular Music," *Popular Music* 22(3) (2003): 337–55.

78. The term *robot*, originally from Czech, was coined as a term for a forced laborer in the play *R.U.R. (Rossum's Universal Robots)* by Karel apek, 1920.

79. McLeod, "Space Oddities."

80. Anderson Ashburn, "Detroit Automation," *Annals of the American Academy of Political and Social Science* 340(1) (1962): 21–28.

81. "Past Products Museum," Roger Linn Design. http://www.rogerlinndesign.com/past-products-museum.html (accessed August 9, 2012).

82. Nate Harrison, *The Amen Break*, 2004. Dubplate recording, playback recorded on video. http://www.youtube.com/watch?v=5SaFTm2bcac (retrieved August 1, 2010).

83. Anonymous, "Cultural Diminuendo."

84. Katz, *Capturing Sound.*

85. Joanna Demers, "Sampling the 1970s in Hip-Hop," *Popular Music* 22(1) (2003): 41–56.

86. Demers, "Sampling the 1970s in Hip-Hop."

87. "Stop Thief," *Melody Maker*, December 19, 1987.

88. Over the last two decades, the open source movement has developed software such as Mozilla Firefox, Arduino, and Pure Data by making the software and its source code free to all.

89. Harrison, *The Amen Break.*

90. Tom Rowlands, e-mail to author.

91. Rowlands, e-mail to author.

92. TyeTye and Defenicial, "Part 1: The Pre-History of Beatboxing." http://www.humanbeatbox.com/history/ (accessed August 1, 2010).

93. *I Love Quincy*, documentary directed by E. Lippmann. USA. Dune Productions, 1984.

Bibliography

Anonymous. "Cultural Diminuendo." *Musicians' Union Monthly Report*, 1959.

__________. "The Fairlight: Computer Musical Instrument. Users Manual." October 1983. Accessed in stores of Science Museum, London (inventory number 1993-654 Pt 11).

Ashburn, Anderson. "Detroit Automation." *Annals of the American Academy of Political and Social Science* 340(1) (1962): 21–28.

Banks, Jack. "Video in the Machine: The Incorporation of Music Video into the Recording Industry." *Popular Music* 16(3) (1997): 293–309.

BBC Sound Maintenance Department. *The Mellotron Sound Effects Console*. London, ca. 1965. In-house manual.

Brend, Mark. *Strange Sounds: Offbeat Instruments and Sonic Experiments in Pop.* San Francisco: Backbeat Books, 2005.

"Closeups of Wurlitzer Sideman." *Deviant Synth*, 2007. http://www.deviantsynth.com/2007/08/23 (accessed August 1, 2010).

Cook, Derek. *How the Mellotron Works.* http://www.echoes-music.co.uk/Mellotron/Start_Here.html (accessed August 1, 2010).

Danzinger, Danny. "Something to Get Hung About." *The Independent*, November 14, 1994.

Demers, Joanna. "Sampling the 1970s in Hip-Hop." *Popular Music* 22(1) (2003): 41–56.

Diva. Directed by Jean-Jacques Beineix. France. United Artists, 1981. Feature film.

"Don't Fall for It." *Musicians' Union Journal*, 1967.

Electric Independence: Moby. Motherboard.TV, 2009. http://www.motherboard.tv/2009/10/19/electric-independence-moby (accessed August 1, 2010).

"Electronic 'Side Man,'" *Popular Mechanics*, November 1960.
Epand, Len. "A Phantom Orchestra at Your Fingertips." *Crawdaddy*, April 1975.
"Farfisa." http://www.combo-organ.com/Farfisa/index.htm (accessed August 1, 2010).
Fink, Robert. "The Story of ORCH5, or, the Classical Ghost in the Hip-Hop Machine." *Popular Music* 24(3) (2005): 339–56.
Flür, Wolfgang. *I Was a Robot.* London: Sanctuary Publishing Ltd., 2000.
Friedman, Matt. *Vintage Synth Explorer.* http://www.vintagesynth.com (accessed August 1, 2010).
Harrison, Nate. *The Amen Break*. Dubplate recording, playback recorded on video. 2004. http://www.youtube.com/watch?v=5SaFTm2bcac (retrieved August 1, 2010).
Hibbert, Stanley. Telephone interview, 1993.
Hinton, Graham. *EMS Electronic Music Studios.* 2002. http://www.ems-synthi.demon.co.uk/ (accessed August 1, 2010).
History of the Birth of Hip-Hop Culture until 1991, Part 1. Rapido Television, ca. 1991. Television documentary. http://www.youtube.com/watch?v=k8IO2KHjXZg (retrieved August 1, 2010).
History of the Musicians' Union. http://www.musiciansunion.org.uk/about-us/ (accessed August 1, 2012).
Holmes, Greg. *The Fairlight CMI.* http://www.ghservices.com/gregh/fairligh/ (accessed August 1, 2010).
I Love Quincy. Directed by E. Lippmann. USA. Dune Productions, 1984. Documentary.
"Instruments Exist Side by Side." *Billboard*, November 6, 1971.
Katz, Mark. *Capturing Sound: How Technology Has Changed Music*. Berkeley: University of California Press, 2010.
Kean, David. *A Mellotron History.* http://www.mellotron.com/history.htm (accessed August 1, 2010).
Kemble & Co. Ltd. *Retail Price List*, 1959.
Keyes, Cheryl L. "At the Crossroads: Rap Music and its African Nexus." *Society for Ethnomusicology* 40(2) (1996): 223–48.
Kraft, James P. *Stage to Studio, Musicians and the Sound Revolution 1890–1950*. Baltimore, MD: The John Hopkins University Press, 1996.
L. G. "The Turn of the Tide." *Musicians' Union Monthly Report*, March 1930. Cartoon.
Martin, Chris. *Streetly Electronics: The Original Mellotron Makers.* http://www.vemia.co.uk/mellotron/ (accessed August 1, 2010).
McLeod, Ken. "Space Oddities: Aliens, Futurism and Meaning in Popular Music." *Popular Music* 22(3) (2003): 337–55.
The Mellatron (sic). British Pathé, 1965. Newsreel. http://www.britishpathe.com/record.php?id=1962 (retrieved August 1, 2010).
"The Mellotron." *Musicians' Union Journal*, November 1966.
"The Mellotron. Instruction Manual." London, Chappell and Co., 1963.
"M.P.P.A. Important Notice. Disputes Pending with Recording Co's and Talking Pictures." *Musicians' Union Monthly Report*, March 1929.
Musicians' Union. "Music While You Work." Exhibition text, 1993.
"The New Wurlitzer SideMan." *Billboard*, May 1960. Advertisement.
"Past Products Museum." Roger Linn Design. http://www.rogerlinndesign.com/past-products-museum.html (accessed August 9, 2012).
"Paul McCartney Plays Mellotron." *Melody Maker*, January 22, 1967.

"Persistent Publicity." *Musicians' Union Monthly Report*, January 1930.
Pinch, Trevor J., and Karin Bijsterveld. 2003. "'Should One Applaud?' Breaches and Boundaries in the Reception of New Technology in Music." *Technology and Culture* 44 (3): 536–59.
Porcello, Thomas. 1991. "The Ethics of Digital Audio-Sampling: Engineers' Discourse." *Popular Music* 10(1): 69–84.
"Questions." *Contemporary Keyboard* Volume 6. GPI Publications, 1980.
Reid, Gordon. "The Rebirth of Cool." *Sound on Sound Magazine*, August 2002.
Robinson, Eric. *Introducing the Mellotron*. Sales brochure, 1965.
Rösing, Helmut. "Listening Behaviour and Musical Preference in the Age of 'Transmitted Music.'" *Popular Music* 4 (1984): 119–49.
Rowlands, Tom. E-mail to author, July 2010.
"Scotch Show Mellotron." *Tape Recording Magazine*, May 1965.
"Stop Thief." *Melody Maker*, December 19, 1987.
Synth Britannia. Directed by Ben Whalley. UK. BBC, 2009. Documentary.
"The Talkie Menace." *Musicians' Union Monthly Report*, ca. 1929.
"Talkies." *Musicians' Union Monthly Report*, August 1929.
"There's a Riot Goin' On." *Super Seventies Rocksite.* http://www.superseventies.com/spslystone.html (accessed August 1, 2010).
TyeTye and Defenicial. "Part 1: The Pre-History of Beatboxing." http://www.humanbeatbox.com/history/ (accessed August 1, 2010).
Various. *Kingdom Come (Arthur Brown)*. Sleevenotes to CD edition, 2003.
Various artists. *The Rime of the Ancient Sampler, The Mellotron Album.* Voiceprint, 1993. CD.
Various forum members. *Discussion on Gearslutz Forum*, May 2010. http://www.gearslutz.com.
"We Lead—Others Follow." *Musicians' Union Monthly Report*, December 1929.

CHAPTER 5

The Magnetic Tape Recorder

Recording Aesthetics in the New Era of Schizophonia

Ragnhild Brøvig-Hanssen

Mechanical reproduction of art changes the reaction of the masses toward art.

—WALTER BENJAMIN[1]

THE INTRODUCTION of the magnetic tape recorder has been described as "a revolution in sound," but it was neither an instance of absolute technological change nor a sudden cultural transformation. It took almost sixty years from the time when Oberlin Smith described the principles of magnetic recording until the company AEG introduced the first successful tape recorder. It was ten more years before it reached the mass market. And for the first few years after the machine arrived in recording studios, it was used sparingly, with skepticism, alongside the already proven technology of the phonographic disc.[2] It also took at least that long for people to realize its true potential in terms of sound quality and multitrack recording. Nonetheless, the magnetic tape recorder did give rise to several dramatic changes. Culturally, it represented a leap forward for the music recording industry: the reduced manufacturing costs with which it was associated led to an overall decentralization of power from the major professional recording companies to small, independent recording studios. During the late 1950s and the early 1960s, hundreds of institutional and private studios arose around the world, and the expansion of regional recording studios (which were far less expensive than the big studios) proved to be decisive for the emergence of youth-oriented rock 'n' roll music in the 1950s. In this chapter, I will, however, focus on how the new recording and editing abilities of the magnetic tape recorder played a key part in changing music itself—its construction and reception but also our conception of what music is.

This chapter falls into three main sections. In the first part I will summarize the development of the artifact itself. The magnetic tape recorder was a great improvement upon the phonograph, in that it offered better sound quality, longer playing time, and a less expensive and more manageable and practical design. This development is not linear but multidimensional and complex, and I will not try to shed new light on the history of the magnetic tape recorder. Instead, I will introduce five distinct models from five important stages in its development. The machines I have chosen are Valdemar Poulsen's Telegraphone (1898), which was the first realization of the ideas of Oberlin Smith; the steel wire recorder, which originated in the 1930s but will be represented here by a Webster-Chicago recorder from the 1950s; AEG's Magnetophon K4 (1938), which is based partly on the ideas of Fritz Pfleumer; the Ampex Model 300 (1948), an upgrade of the Model 200, which was itself built on the technology of the Magnetophon K4; and, finally, Ampex's eight-track Sel-Sync recorder (1956), which was inspired by Les Paul's overdub experiments. I will discuss the motivations behind the development of these machines and their individual relationships to the magnetic tape recorder's signature achievement: a completely new era of what Canadian composer and writer R. Murray Schafer has labeled *schizophonia*; that is, a new level of spatiotemporal disjunction of sound.

This spatiotemporal disjunction will then occupy the next part of the chapter. Before the invention of the phonograph in 1877, music could only be heard according to the acoustic laws that applied to its live performance. The phonograph, on the other hand, introduced a distinction between original and reproduced sounds, hence Schafer's term (*schizo* is "split" and *phonia* is "sound" in Greek).[3] The magnetic tape recorder, with its multitracking ability, led us into yet another era of schizophonia by extending the temporal and spatial parameters of the music it recorded. Recorded music came to imply a patchwork of sounds recorded at different times and in different spaces. The magnetic tape recorder changed the standards for producing and composing music, but it also impacted the reception of recorded music, which could no longer be trusted to be a faithful reproduction of a preexisting and coherent event.

In the third part of the chapter I will discuss how musicians and sound engineers applied the magnetic tape recorder's new recording and editing abilities in very different ways. While some used this recording equipment in a way that was similar to its predecessors, others embraced the changes it implied. I have identified three distinct recording paradigms that emerged in the wake of this artifact's ascent to everyday use, and that remain perfectly valid today: the "documentary event," the "ideal event" and the "surrealistic event." I will describe each paradigm in turn using early examples of how the new medium was approached and demonstrating its relevance to each one.

The Development of the Magnetic Tape Recorder

In 1928 the Australian inventor Fritz Pfleumer glued pulverized iron particles onto coated cigarette paper and installed the paper on a self-made reel-to-reel machine. This machine of "sounding paper," as he called it, was in fact the first magnetic tape recorder, although the sound quality was too poor for the machine to be of any use.[4] However, just as the phonograph is often wrongly attributed entirely to Thomas Edison, the magnetic tape recorder was less the brainchild of Pfleumer than an outcome of a whole constellation of social, economic, and cultural factors, as well as a whole series of previous experiments, discoveries, and inventions. American engineer Oberlin Smith identified the basic principle behind magnetic recording as early as 1878, the year after Edison patented the phonograph, as he tried to minimize the latter's background noise. After a visit to Edison's studio, Smith started to think about how to improve the recording device. He began with the technology of the phonograph, which worked as follows: a recording horn captures and concentrates the performed sounds, and a diaphragm (membrane) placed at the end of the horn vibrates in response to the sound waves (like the ear's tympanic membrane). A stylus (pointed tool) that is connected to the diaphragm moves in line with the diaphragm's vibrations, cutting a groove similar to the vibration patterns of the sound waves into a cylinder or disc. This mechanism can be seen at the photography of an Edison Home Phonograph from ca. 1910, which is today part of the Norwegian Museum of Science and Technology's collection (Figure 5.1).

The phonograph was therefore mechanical in nature, in the sense that it relied upon the conversion of one motion into another via physical contact. That contact between stylus and cylinder (or disc),[5] however, resulted in significant background noise that in turn compelled Smith to explore a recording method that was not mechanical. In the interests of improving the sound quality of the recorded material, Smith turned to the technology of telephony, in which sounds were transformed into electric currents. He described his notion of "magnetic" recording in a letter he wrote the same year as his visit to Edison's studio: "While talking into a mouthpiece, the varying intensity of current . . . produces zones, or spots, of magnetism in the wire which vary in length and strength in accordance with the length and amplitude of the sound vibrations."[6] Smith rightly predicted that when electric currents converted from sounds are placed near a magnetizable material such as a wire, a magnetic field on the wire would emerge, and the patterns of the currents would produce a similar magnetic pattern on the wire. In this way sounds could be stored on a medium without physical contact. The first-known realization of this idea arrived twenty years later, when the Danish engineer Valdemar Poulsen demonstrated and patented his Telegraphone—an electromechanical recording medium, initially invented for office dictation and recording telephone conversations. The Telegraphone shown in Figure 5.2 was manufactured by the Danish development and production company Dansk Telegrafonfabrik, which was established three years after Paulsen had demonstrated his machine at the 1900 Paris Exposition. Although the

FIGURE 5.1
An Edison Home Phonograph (ca. 1910). Courtesy of the Norwegian Museum of Science and Technology.

Telegraphone showed great promise and was by some regarded "the next phonograph," the company went out of business in 1917 (just like the American Telegraphone Company, 1903–1919). One of the reasons for this was manufacturing problems; another was poor sound quality. Despite the fact that the recorded message had better clarity than the phonograph, the volume of the recorded sounds of the Telegraphone was very low, and the medium itself had to await the development of adequate amplifiers in order to become commercially successful.[7]

Improvements in electronic amplification eventually led to the commercial establishment of steel tape and wire recorders during the 1930s (as well as a shift within phonograph technology already in the mid-1920s from mechanical to electromechanical recording).[8] The solid-steel recorder had significantly longer recording time than the three minutes permitted by the phonograph, and its robust design was more fit for traveling as well. These advantages made it particularly popular as a medium for dictation, and during the war it was used for military operations, such as secret audio surveillance.

FIGURE 5.2
Valdemar Poulsen's Telegraphone, produced by Dansk Telegrafonfabrik in Copenhagen between 1903 and 1917. Courtesy of Danmarks Tekniske Museum.

Between 1945 and 1955, several companies in the United States, South America, and Europe produced relatively low-cost wire recorders that reached a new consumer market of individuals, companies, and institutions (such as home recording hobbyists, composers and musicians, businessmen, teachers, and historians). Despite Smith's original motivation to improve the sound quality of recorded material, and despite improvements in electronic amplification, the volume of the sound signal of steel-wire magnetic recorders stayed significantly lower than that of the phonograph. The Webster-Chicago steel tape recorder Model 181 from ca. 1948, which is also represented in the collection of the Norwegian Museum of Science and Technology (Figure 5.3), was promoted as "the studio model," but the phonograph[9] remained the standard medium in recording studios until the magnetic tape recorder could finally outdo it in the 1950s.[10]

Pfleumer's actual motivations for switching from standard solid-steel material to the aforementioned softer, nonmagnetic coated material (in this case, cigarette paper) with iron particles are unclear, but the advantages were several. Tape was less expensive and

FIGURE 5.3
A Webster-Chicago steel tape recorder Model 181 (ca. 1948). Courtesy of the Norwegian Museum of Science and Technology.

easier to handle than wire, it did not tangle while rolling, and it provided better sound quality. In 1932, Pfleumer signed a contract with the German electronics group Allgemeine Elektricitäts-Gesellschaft (AEG), which recognized the market for improved magnetic recorders and believed in his idea. AEG, in collaboration with the giant iron carbon manufacturer I. G. Farbenindustrie Aktiengesellschaft, devoted considerable resources to developing Pfleumer's invention. Hoping to improve problems such as the tape's poor resistance to tearing, short playing time, disturbing background noise, and poor sound quality, they experimented with different tapes and magnetizable materials, tape widths, and speed of tape, and they developed different types of transducer heads.[11] In 1938, AEG succeeded in introducing the first commercially viable tape recorder: the Magnetophon K4.[12]

The sound quality of the Magnetophon K4 was significantly better than that produced through wire reproduction, and it played longer than the phonographic disc. It was also less expensive to produce and easier to maneuver than either the contemporary phonographic disc player or the steel wire recorder. Nevertheless, during the war, at least, Magnetophons were produced only for military needs.[13] For instance, the K4 that today is part of the Norwegian Museum of Science and Technology (Figure 5.4) was used in Norway for military purposes. After the fall of Nazi Germany, the U.S. Alien Property Custodian took charge of all of the patents on the machine.[14] The American electronics company Ampex decided to develop its own tape recorder based on the technology of the Magnetophon K4 and secured financing from American singer Bing Crosby, who was eager to find a method to prerecord his radio shows with better sound quality than phonographic discs could offer. The following year the company introduced the Ampex Model 200, the first successful American tape recorder, and it proved to be crucial to the eventual establishment of magnetic tape as the standard recording medium in broadcasting and recording studios of the era.[15] As competitors quickly emerged, Ampex developed a less costly design: the Model 300, first produced in 1948 (see Figure 5.5 for an Ampex Model 300 from the Canada Science and Technology Museum). Improvements in tape material and tape heads allowed the transport speed to be halved, and, consequently, the reels could be reduced in size without affecting the recording time. Because of this, the Model 300 offered the same quality as the Model 200, despite being smaller in size and less expensive to produce; its design served as a lasting model for many portable recording machines to come.[16] By the mid-1950s, the magnetic tape recorder was in general use throughout the sound-reproduction industry, and over the next thirty years, manufacturers worldwide would continue to improve the technology, focusing on ever cheaper and smaller machines that would produce less noise and be easier to operate.[17]

Of course, consumers often use machines in ways that are not anticipated by their manufacturers, who might then choose to develop new machines based on those creative applications. This was the case with the development of the multitrack recorder. With the phonograph, it was only possible to record one track at a time, so musical performances were usually captured as a unit and any possibilities for further editing of the recorded sounds were limited. The American guitarist and inventor Les Paul developed an overdub (or sound-on-sound) technique by modifying his Ampex Model 300 with an extra playback head and altered tape path, but the manager of Ampex's Special Product Section, Ross H. Snyder, saw the need for a technology that would improve this process of overdubbing even more. Though multichannel recorders had existed for some time, their channels could only record simultaneously, in real time, and stored on one track only. Snyder therefore invented a technology called Sel-Sync (a shortening of "selective, successive, synchronous") that made it possible to record individual tracks separately and in complete isolation from one other.[18] Parts could now be recorded separately at different times, and, if desired, in different locations. Also, because sounds could be recorded through several channels without being automatically bounced onto a single

FIGURE 5.4
Magnetophon K4, which was AEG's first commercially successful tape recorder. The machine in this photograph dates back to 1939, the year after the first machine was introduced. It was used in Norway during World War II for military purposes. Courtesy of the Norwegian Museum of Science and Technology.

FIGURE 5.5
The Ampex Model 300. While the Ampex Model 200 was the first successful U.S. tape recorder, the design of the Ampex Model 300 served as a model for many future portable recorders. Courtesy of the Canada Science and Technology Museum. Photo: Peter Lindell.

track afterward, the tracks could be treated separately even after they had been recorded. Ampex built its first eight-track Sel-Sync recorder in 1956 (and sold it to Les Paul the next year for $10,000). The multitrack recorder was a huge step forward in the overdub technique, which had required musicians to erase all of the old tracks if just one overdub take was unsuccessful; it also solved the problem of degradation in sound quality that took place after each new overdub. Snyder recalled his modest initial hopes for the invention, and the mistaken assumptions about its limited application: "[I wanted to] improve the recording process for those doing overdubs for any reason . . . I mistakenly thought its usefulness somewhat narrow, and did not dispute Ampex's patent attorneys' advice that it might be taken as obvious art, thus probably not patentable."[19] Fate would prove otherwise for the new technology. The praxis of constructing music out of several takes soon became the standard way of making recordings, whatever one's methods and ultimate musical goals.

The primary motivation behind the development of the magnetic tape recorder, from the invention of Oberlin Smith to the successes of Ampex, had been to improve the sound quality and playing time of recording machines while developing ever cheaper and more practical designs. More than merely improving the technology of recording, the magnetic tape recorder set in motion two significant shifts in the history of recorded music. First, it allowed for a new level of spatiotemporal disjunction in sound, thus leading us into a new era of schizophonia. Second, its new recording and editing possibilities—and consumers' various applications of them—played a significant role in the establishment of the three recording paradigms that still dominate the field today. In short, the magnetic tape recorder changed our whole conception of music, and it is that conception to which I will turn next.

A New Sense of Time and Space

The invention of the phonograph in 1877 challenged our traditional understanding of sounds as emerging directly from a live source. As mentioned, Schafer therefore points to this invention as the dawning of the era of schizophonia. In his important essay "A Voice without a Face" (1991), Dave Laing points to the fact that members of the audience fainted when Edison demonstrated his speaking phonograph in 1888, reminding us that this disembodied sound "must have been a vital shift in the experience of listening to music."[20] Early print advertisements from the recording industry play upon this point: the iconic RCA Victor dog Nipper sits alertly in front of a recording horn, curiously listening to "His Master's Voice," while an Edison Company ad depicts a child destroying a phonograph while "Looking for the Band."[21] While the already familiar telephone also mediated a voice without a face, those sounds remained "live"—that is, they were produced at the same time that they were being heard. This incidence of *spatial* detachment paled in comparison to the phonograph's *temporal* detachment, whereby the reproduction of sounds did not enjoy a necessary relation to a simultaneous, if distant, source.

However, although the sounds of a musical performance were cut loose from their origin in time and space, they nevertheless remained a unit, in the sense that what you heard from the recording was the sound of a preexisting coherent event that had been recorded in one take.

The rare exceptions to this were recordings that resulted from very early applications of the technique of overdubbing. With two separate recording machines, one could record a machine playing already recorded material while recording new sounds atop it. Thus, the final product would reproduce the sounds of a combination of two or more different events rather than a single, coherent event. Although overdubbing was used occasionally with mechanical and electromechanical recording, the phonograph was generally viewed as an archival medium rather than a creative tool; Edison himself celebrated its ability to "preserve and hear again . . . a memorable speech, a worthy singer . . . the last words of a dying man . . . of a distant parent, a lover, a mistress."[22] Advertisements promoted recordings as "lifelike," a "true mirror of sound," "natural" or "the real thing,"[23] and early musical recordings were similarly promoted as archived events, or copies of original performances. Given this context, Jason Toynbee reads the slow adoption of sound-manipulation techniques less as a result of their perceived lack of promise than as a by-product of the conservative approach to new possibilities around framing performances, which he sees as characteristic of the history of music and technology in general.[24] Whatever the larger cultural bias, experiments were undertaken and significant technical difficulties were encountered. Thus, although the possibility of manipulating time and space in the reproduction of music had existed since the birth of the recording medium, it was only through the magnetic tape recorder and the invention of multitracking that it became truly viable. Thus, John Philip Sousa's slur on phonograph music as "canned music"[25] is in fact an accurate description of how the recording medium was used in its early days; the event represented by the mechanical recording medium remained "trustworthy," in that the recorded sounds could be traced directly to musicians who were playing their instruments at the same time and in the same space.[26]

While the invention of the phonograph represents a shift to schizophonia, then the invention of the magnetic tape recorder brought about a new era in it, given the dramatic new possibilities for spatial and temporal disjuncture between sound and its source(s). Sounds could be thoroughly detached from their spatiotemporal origins *and* juxtaposed with other sounds with other origins. For instance, tape made it possible to literally cut tracks apart and paste them together again through the process of splicing: after placing the sound sequence that needed to be split between the two open reels of the tape recorder, the engineer used a ruler to locate the exact spot and then cut the tape with a razor blade, guided by the channel of a metal splicing block. Using a piece of editing tape, he or she could then recombine the two loose ends into a continuous sound section (Figure 5.6). Through splicing, engineers were able to freely juxtapose musical tracks from different times and places, and the fact that the music represented such a patchwork was not always even audible as such. Later in this chapter I will review the various ways

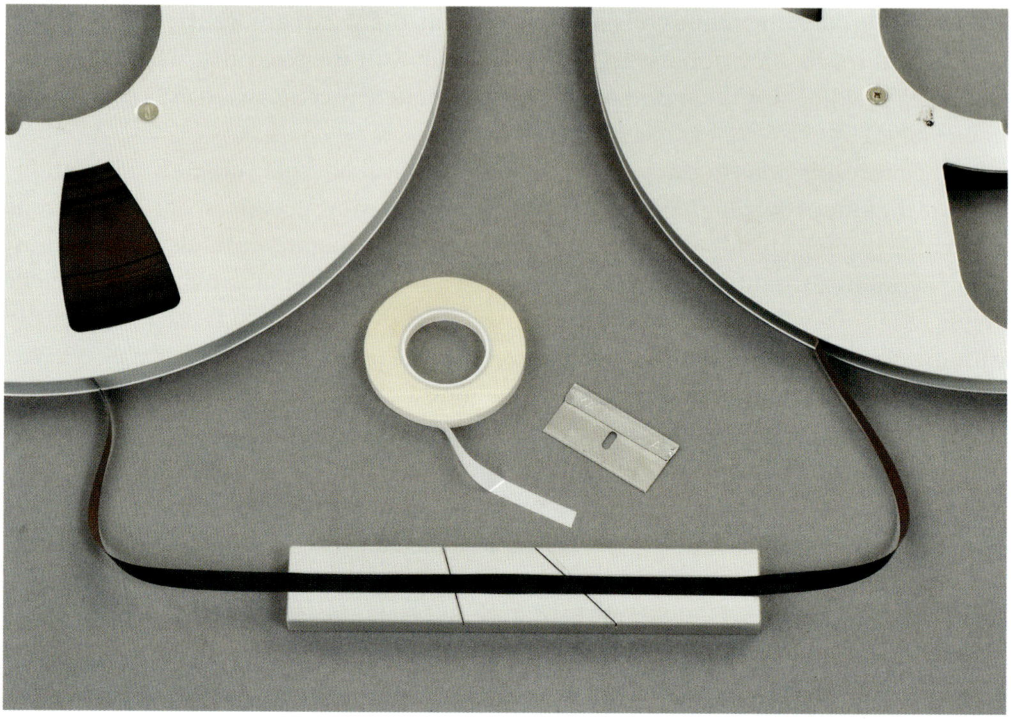

FIGURE 5.6
The photograph shows the process of analog tape splicing. The tape is placed in a splicing block and cut with a razor blade guided by the splicing block's channel. Editing tape may then be used to merge the spliced tape ends. Courtesy of the Norwegian Museum of Science and Technology.

engineers approached splicing, and especially their early experiments with audible cuts, tape loops, and tape delay.

Multitrack recording also facilitated the possibility of sonically challenging our traditional sense of time and space, since the engineer could record several individual takes at different times as well as edit the tracks separately. By deliberately placing individual sounds within the stereo field[27] and/or adding processing effects such as reverb and echo to them, the engineer could mold whole virtual environments, even those that appear contrary to natural acoustic laws. These recorded performances either never actually existed but sounded real, or they never actually existed and they did *not* sound real, thanks to their transcendence of the spatiotemporal laws of acoustics. I will discuss these possibilities further in the third part of this chapter. This new capacity for manipulation altered the way musicians, producers, and sound engineers worked in the studio while recording and mixing sound. Certain pioneers began to regard the studio less as an archiving center than as a laboratory for "sculpting" patchwork performances out of several takes. Listeners soon approached recorded music differently as well—after all, these sounds were no longer "trustworthy" in terms of their fidelity to an original temporal and spatial form.

In this new era of schizophonia, fiction could easily be masked as truth. Because of its new recording and editing abilities, then, the recording device had transitioned from an archival medium that *preserved* musical events to an artistic medium, one that represents events with no claim to their preexistence.

The Establishment of Different Recording Paradigms

A profound enrichment in recorded musical expression resulted from this artifact's increased capacity for manipulating sounds. Some engineers embraced the new technology and experimented with the new recording and editing abilities in entirely unanticipated ways, even as others continued to apply it to the task of archiving actual performances. I have identified three distinct recording paradigms that emerged from this cultural and industrial ferment: the "documentary event," the "ideal event," and the "surrealistic event." Although these three categories must be seen as analytical classifications of musical ideals that are in practice not always so clear-cut, they remain perfectly valid today. I will describe each paradigm in turn and discuss the essentially different core values upon which each is based.

The Documentary Event

In the new era of schizophonia, music makers (musicians, producers, and sound engineers) suddenly got a whole new set of tools with which to create music. However, not everyone availed themselves of their capacity to manipulate sound; some continued to view the recording machine as exclusively a transmitting medium whose main function was to archive coherent musical performances. This reticence centered upon any technique that might disrupt the recording's faithfulness to a preexisting musical event, and to some this included the ability to splice the tape (this despite the fact that experimentation with splicing was already taking place with optical film in the motion picture industry). Musicians deplored the technique for ruining the spirit of the music. For instance, Russian-born classical cellist Gregor Piatigorsky (1903–1976) said, "I don't like splices, I don't like any falsehood . . . I don't like any perfection . . . If the spirit is there, it's good enough for me."[28] The technique challenged the musician's belief in spontaneity and improvisation, which was, according to Evan Eisenberg, why splicing met resistance within the field of jazz music: "When splicing did become a possibility, jazz musicians resisted it; as improvisers they believed, even more passionately than old-fashioned classical musicians, in the spontaneity of the long take."[29] For instance, Miles Davis's legendary album *Kind of Blue* (1959) consists, according to his pianist Bill Evans, only of first takes, and when he was forced to use multiple takes in other situations, it was "without calm."[30] While some musicians and engineers stuck with the more conservative approach, others accepted small alterations, like using the technique of splicing in a discreet or entirely hidden way to eliminate only unwanted sounds or to move a sequence from one take to another in order to make things sound better. They thus combined the tape ends in a silent spot in the performance or used various techniques (such as the "hourglass" splice) to make the joins smooth and inaudible. At stake in my "documentary" paradigm, then, is not an *un*mediated performance but the claim of *truth*: the recording must be able to be taken as a faithful signifier of a preexisting event (despite small alterations).[31] In this sense, my use of the term evokes its filmic application. Paul Arthur reviews a variety of strategies within documentary film but points out that they are all "wedded to the same principles of authenticity, if not the same rhetorical codings"[32]—each asserts its truth.

Today, recordings in which subtle editing is accepted likely outnumber strictly conservative recordings in which nearly all forms of technological manipulation are rejected. Yet, the ideal of the "documentary" paradigm persists. This is obvious from the many CD covers, photographs, and music videos depicting artists in a live setting (even artists who seldom, or never, do live performances)—they imply a performance that has happened, for which the recording or music video is merely intended to substitute.[33] Such reality-driven recordings ultimately seek to reconstruct, or at least suggest, what Walter Benjamin calls the "aura" of a performance, or "work"—that is, the performance's "presence in time and space, its unique existence at the place where it happens to be."[34] This aura might include the sound of musicians counting off before the music starts or talking to each other during the performance—both of which signify in no uncertain terms the reality of the recording session. Similarly, coincidental coughing, calling, or laughing, as well as musical "errors," usually indicate a single recorded take rather than an idealized amalgam of several takes. Eisenberg reviews other examples from classical recordings, such as the sound of church bells at the start of a performance that was not even recorded in a church.[35] Ironically, then, successful sound manipulation within this ideal in fact enhances the impression of *no* manipulation: there are examples of recordings in which applause is *added* to a studio performance to make it sound like a live event. Thus the "documentary" paradigm implies the urge to be real as much as the actual achievement of it. Michael Renov's description of documentary film as "the creative treatment of actuality"[36] also applies to the "documentary" paradigm of music. By reconstructing qualities that are often associated with the purportedly preexisting performance, the recording presents itself as being documentary, though inevitably with a decayed sense of presentness (what Benjamin calls a "decayed aura"), so that the listener feels as if he or she is *almost* witnessing it in the first person.

Some might argue, then, that this paradigm did not *arise* in the wake of the magnetic tape recorder but instead pre-dated it as the traditional approach to the recording medium. Certainly the recording medium had a transmitting function that dates back to the phonograph, and most recordings were already documentary in the sense that they were documentations of musical performances that had taken place in the real world. However, this was less a choice than a necessity, because the available technology made it extremely difficult to record in any other way. With the magnetic multitrack recorder, however, the documentary approach soon became an aesthetic choice in an era of schizophonia that boasted a range of new manipulation possibilities and (at least two) alternative paradigms.

The Ideal Event

Often, scholars contrast the paradigm that I have called the documentary event directly with avant-garde techniques or experimental music (which in most cases fall under my paradigm of surrealistic events). There is, however, also a middle position, which I have called the "ideal event." Unlike those with documentary priorities, some engineers and producers used the new recording and editing capabilities to create ideal events *without* claiming that the represented event actually happened. By way of emphasizing the gap between the paradigms of the documentary and the ideal event, I would note Eisenberg's reference to a relatively recent advertisement for a particular brand of compact disc in comparison to the aforementioned ads for the phonograph ("lifelike," "a true mirror of sound," "natural," and "the real thing"): "How would you like to hear music at home the way the engineer hears it in the studio? The Magnavox Compact Disc . . . It's like being in a recording studio."[37] To Eisenberg, this ad demonstrates that "the ideal is no longer live music, but some technologic Platonic form."[38] According to the paradigm of the ideal event, then, it is the *idea* that matters—the sonic result alone—rather than its preexistence in "real life."

An early example of this paradigm appears in the recordings of Glenn Gould (1932–1982), the Canadian pianist who has been described as "a passionate champion of splicing"[39] and "a pianist wedded to the record."[40] Gould was delighted by the ability to make recordings that were free from flaws and mistakes, and he promptly went against the existing ideology by constructing "perfect" performances out of chopped-up sequences from multiple takes. Unlike those who secretly used the tool to correct minor mistakes, Gould made no attempt to deceive his listeners about his devotion to virtual perfection: in interviews he was always eager to talk about his creative process, and at the age of thirty-one he stopped performing live and dedicated himself to recording alone.[41]

While recordings within the paradigm of the documentary event claim to represent preexisting events from "real life" (events that purportedly actually did happen), recordings of ideal events represent, as the name implies, events from the world of ideas, constructed for the recording only. Nevertheless, both paradigms treat the recording medium as a window to imaginable continuous performances, whether actual or virtual. This distinguishes them from "surrealistic events," which reveal themselves as montages of several different events.

The Surrealistic Event

While some music makers stuck with using the magnetic tape recorder just like the phonograph, and others used its new editing possibilities to create ideal (but *potential*) events, still others took a more experimental approach altogether, abandoning the notion of presenting a recording of an event-based performance. For instance, participants in the early-1950s electroacoustic music scene pioneered techniques with magnetic tape that fragmented spatiotemporal structure so that the recording medium itself became "opaque" instead of "transparent."[42] The magnetic tape recorder in this case was no longer a transparent window to a coherent event but an enabler of something else: the surrealistic event. It is perhaps problematic to oppose "surrealistic" with "natural" soundscapes in music, because the notion of the "natural" is particularly fluid. In the context of this chapter, however, the term *surrealistic* will identify those recorded musical expressions that have no immediate allegiance to a performance of any sort, actual or virtual. In a physical environment, of course, we cannot juxtapose different times and different spaces; at best, we might simulate the effect. It is this concreteness, and its attendant limitations, that I set against the freedoms of the recorded surreal.

By pushing the new medium to its limits, these "musical surrealists" separated sounds from their temporal and spatial origins in a way that made obvious the resulting music's fragmented construction. Tape splicing was, for instance, soon harnessed to the production of unique sonic effects. In 1951, the American composer John Cage organized the Project of Music for Magnetic Tape, whose aim was to explore tape as a medium for creating music itself, and in 1952 he and other participants in the project composed *Williams Mix*, a juxtaposition of hundreds of spliced tapes.[43] According to Cage, "The chief technical contribution of my work with tape is in the method of splicing, that is, of cutting the material in a way that affects the attack and decay of sounds recorded."[44] A sound, after all, consists of an "attack" (the sound's onset), a "steady state" (the middle section of a sound), and a "decay" (the sound's fadeout), but this acoustic process can be artfully disrupted by splicing. This technique was also used by the French pioneer of *musique concrète*, Pierre Schaeffer, who, in his project to facilitate "reduced listening" (the act of listening to sound for its own sake), attempted to make sounds unrecognizable in order to blot out any potential associations with their sources. Already known for manipulating sounds by, for example, altering the speed of the phonograph's turntables, he started experimenting with tape in the late 1940s and soon discovered that by physically cutting off part of a sound's attack, it became much less recognizable.[45] In 1952, the young German composer Karlheinz Stockhausen regularly visited Schaeffer's *Groupe de Recherces Musicale* (GRM)[46] studio in Paris for instruction in the art of tape editing, and he recalls experiments involving the insertion of leader tape (blank, nonmagnetic tape normally used at the beginning and ending of a recorded song) between sounds to create a percussive, stuttering effect.[47] Stockhausen also created continuously sustained sounds using the technique of tape looping: "I copied each sound many times and, with scissors, cut off the attack of each sound. A few centimetres of the continuation [remaining sound], which was, briefly, quite steady dynamically, were used. Several of these pieces were spliced together to form a tape loop."[48] One creates tape loops—that is, sequences of sound repeated over and over—by pasting both ends of a spliced sequence together and then placing the loop on a tape recorder for playback. While loops could be created using the phonographic disc—for instance, Schaeffer created locked grooves with a disc cutter to repeat the sounds[49]—magnetic tape made the operation much more straightforward (and, soon enough, more common as well).

According to Thom Holmes, the first electronic music experiments in the 1950s only existed on tape and were never performed live.[50] By the mid-1960s, however, the staging of live performances of electroacoustic music had become more widespread, partly due to the increased availability of portable tape recorders, amplification systems, and synthesizers. Gentle Fire was a leading performance group from this first era of electroacoustic live music.[51] The group combined live performances on traditional and invented instruments with prerecorded tape material, which they produced using the techniques pioneered by Schaeffer, Cage, and Stockhausen, among others.[52] One such technique involved altering the sonic characteristics of the recorded material by speeding up or slowing down the pace of a strip of tape during playback. When relatively sustained chords are sped up, they may suddenly realize a pronounced rhythm; conversely, when a pronounced rhythm is slowed down, its components may appear to be sustained chords. Tape recorders at this time were usually equipped with only one switch for two different speed options, but they could be modified to offer a continuous "varispeed" option as well. The Science Museum in London possesses a Revox A77 that was adjusted by Gentle Fire member Hugh Davies, who was particularly renowned for his self-invented instruments or "sounding objects" (Figure 5.7). The group also favored backward-playing—the playback of recorded sounds in reverse, either by chopping out a sequence in the tape and replacing it but backward, or turning the tape over and running it backward behind the playback head. Backward playing had also been experimented with in early phonographic avant-garde music, but, as with looping, tape made the whole process much more straightforward. Lastly, the group experimented with the generation of tape delay or echo, in which a sound could be repeated once or several times. By adding an extra playback head to the recording machine and combining a reel-to-reel tape with a looped tape sequence, the signal on the reel-to-reel tape would play back while being recorded on the tape loop. When the tape loop ran through the playback head a few seconds later, the sound that was just heard was repeated; the tape loop then entered the erase head to begin the process again (the length of the tape path from the recording head to the playback head determined the delay time). Several of Gentle Fire's tape loops are today part of the Hugh Davies collection at the Science Museum in London (Figure 5.8). These tapes were probably used both to create delay effects and, like a modern digital sampler, to repeat snippets of sounds consecutively.

These early experimental techniques were soon adapted to the field of popular music. Sam Phillips pioneered the "slapback"—a type of echo with a short delay time and only one repetition—at Sun Studio in 1953 or 1954, first in a recording by the American blues musician Doctor Ross but most famously in some early recordings by Elvis Presley. Phillips achieved the echo effect by bouncing the sound signal between two Ampex recorders.

FIGURE 5.7
Varispeed Revox A77 tape recorder used by electroacoustic musician Hugh Davies. Courtesy of the Science Museum (Inv: 2007-127). Images available from Science & Society Picture Library.

Peter Doyle notes with regard to these recordings that "there seems to be little attempt here to create a consistent, believable spatiality,"[53] and Toynbee describes the slapback in Elvis's Sun recording of "Mystery Train" (1953) in a similar way: "The place that we are taken to as we listen is emphatically not a concert hall, bar, or lounge though. Rather this is a virtual architecture, one that is much 'larger than life.'"[54] In other words, like the reversal of sound, experimental cut-ups, and tape loops, the slapback implied a surrealistic environment and musical event rather than a preexisting coherent performance. Likewise The Beatles' recordings at Abbey Road Studios in the 1960s also exemplify early experimental approaches within the popular music field, especially their groundbreaking albums *Revolver* (1966) and *Sgt. Pepper's Lonely Hearts Club Band* (1967). For instance, "Tomorrow Never Knows" from *Revolver*, produced by George Martin and engineered by Geoff Emerick, contains a number of tape loops. Martin described the process of recording the loops: "All over the studio we had people spooling them [the loops] onto machines with pencils while Geoff did the balancing"; Emerick adds: "I laid all the loops onto the multi-track [of five machines] and played the faders like a modern day synthesiser."[55] On "A Day in the Life" from *Sgt. Pepper's Lonely Hearts Club Band*, the producers

FIGURE 5.8
Assorted tape loops used by members of Gentle Fire in their electroacoustic musical performances. Courtesy of the Science Museum (Inv: 2007-137). Images available from Science & Society Picture Library.

manually synchronized two four-track Studer J37 tape recorders (Figure 5.9), so that the rhythm tracks of The Beatles could be recorded on one machine while an orchestra of forty musicians was recorded four times, on all four tracks, of the other machine. As a result, those forty musicians sounded like four times as many backing the band.[56]

Within the construction of these surrealistic events, "space and time became equivalent forces to be worked like any other material substance of music," writes Holmes. In a way, the tape recorder even *materializes* space and time: "Holding a strip of tape in your hand was like seeing and touching sound."[57] This experimental paradigm was, however—like the paradigm of the ideal event—not exclusively the result of new technology; the phonographic disc recorder had also been used in a "surrealistic" way. For instance, in 1939 John Cage premiered his *Imaginary Landscape No. 1*, a composition for muted piano, cymbal, and two phonographs in which two artists varied the speed of their sound discs at the turntables. As Holmes puts it: "Composers had been waiting for a device that would allow them to store and manipulate sounds better than the acetate disc. When the magnetic tape recorder came out of postwar Germany, they knew exactly what to do with it."[58] However, the magnetic tape recorder, with its ability to record the

FIGURE 5.9
Studer J37, identical to the two machines used by The Beatles. Courtesy of the Norwegian Museum of Science and Technology.

sounds at different times and in different surroundings and to further manipulate them after recording, facilitated a far greater array of techniques that in turn propelled this aesthetic forward. Holmes points out that while many found the experimental music of the electroacoustic music scene "a bit too radical for their taste," the influence of this paradigm "began to broaden their opinion about what was and was not musical."[59]

Conclusion

We have seen how Oberlin Smith's early notion of magnetic recording from 1878 was motivated by a desire to minimize the noise caused by the mechanical technology of the phonograph, and how Valdemar Poulsen realized this notion in his Telegraphone twenty years later, despite the machine's ultimately inadequate sound quality. Electronic amplification would later enable a more successful realization of Smith's original idea, but the sound quality would remain poor until the advent of AEG's Magnetophon K4, produced a full sixty years later. The Magnetophon also had other advantages, such as a more affordable, robust, and manageable design than its predecessors (and especially the phonograph). Ampex sought to re-create the Magnetophon K4 in America and ended up with a machine that was even smaller and less expensive—the Ampex Model 300 from 1948. This machine was further developed into the first successful multitrack recorder in 1956—Ampex's eight-track Sel-Sync recorder was designed to supersede the still relatively rare and awkward practice of overdubbing. It succeeded, and soon such practices became the standard means of producing music. In this chapter, however, I have argued that the development of the magnetic tape recorder represents much more than a step forward (however great) in the technical development of recording machines.

I borrowed Schafer's term *schizophonia* to describe the spatiotemporal break initiated by the phonograph, whereby sounds for the first time were detached from their sources in space and time. I then argued that the magnetic tape recorder in fact brought about a whole new era of schizophonia. If phonograph recordings were mainly "canned" or "hermetic" music, in the sense that the sounds of a musical performance were at least detached from their sources as a coherent unit, the new recording and editing possibilities offered by the multitrack magnetic tape recorder made it possible to create patchwork performances out of sound fragments that might well have occurred at different times and in different spaces. Because of this potential for illusion, recorded sounds could no longer be trusted to be honest or authentic signs of something that had happened in the way it was heard. Consequently, the recording machine went from being a medium that preserved former events to a medium that could only represent them with no guarantee of veracity.

I then demonstrated how these new recording and editing abilities were actualized in very different ways during what proved to be a very creative time. While some used the recording equipment of magnetic tape in a way that was similar to its predecessors, others saw new possibilities for improving upon reality, and still others embraced the changes implied by the new medium, experimenting with its capacity for fragmented construction. Seen in retrospect, these different approaches led to the establishment of three alternative paradigms for recording that I have called the "documentary event," the "ideal event," and the "surrealistic event." These paradigms remain relevant even today. I described the paradigm of the "documentary" event as interested in representing performances in such a way that listeners are convinced that what they hear once existed as such in "real life." The paradigm of the "ideal event" aims to present performances that theoretically could have happened but need not have—contrary to the documentary, devotees of the virtual will happily sacrifice veracity for perfection. The recording paradigm of the "surrealistic event" diverges from the other two in that it foregrounds the recording's potential for extreme temporal and spatial fragmentation. It is surrealistic in the sense that the musical "event" could not be realized in a natural physical environment without the aid of studio technology.

As mentioned, surrealistic musical events had existed in the era of the phonograph, and the archive-based phonographic recording approach still exists today. Thus, none of the paradigms represents a strictly logical derivation of the technology of the magnetic tape recorder. Andrew Feenberg explains: "Technical development does not point definitively toward any particular path. Instead, it opens branches, and the final determination of the 'right' branch is not within the competence of engineering, because it is simply not inscribed in the nature of the technology."[60] However, it is clear that some of these paths or branches regarding the use of the magnetic tape recorder would not have come about were it not for the technical nature of the machine. The magnetic tape recorder introduced new possibilities for editing and recording that made it possible to make musical recordings in a completely new way—one could sonically create a world that diverged from our everyday life—which led in turn to an enhancement of the ability to express oneself musically through recording. Certainly some resisted experimental music because it abandoned the recording's documentary status as well as the traditional event-based musical ideal. More and more listeners, however, began to join in celebration of the spatiotemporal disjuncture of recorded sounds. Soon, musical recordings were regarded as an independent form of art.

Notes

1. W. Benjamin, "The Work of Art in the Age of Mechanical Reproduction," in *Illuminations: Essays and Reflections*, ed. H. Arendt (New York: Schocken Books, 1968 [1936]), 234. See also W. Benjamin, "The Work of Art in the Age of Its Technological Reproducibility," in *The Work of Art in the Age of Its Technological Reproducibility and Other Writings in Media*, ed. M. W. Jennings, B. D. and T. Y. Levin (Cambridge, MA: The Belknap Press of Harvard University Press, 2008 [1989]), 36.

2. See A. Millard, *America on Record: A History of Recorded Sound* (Cambridge, UK: Cambridge University Press, 1995), 207.

3. Regarding Schafer's work on "schizophonia," Steven Feld rightly observes that the elder scholar had a "suspiciously anxious view of the impact of technology on musical practices and sound

environments" (S. Feld, "From Schizophonia to Schismogenesis: On the Discourses and Commodification Practices of 'World Music' and 'World Beat,'" in *Music Grooves: Essays and Prologues*, ed. C. Keil and S. Feld [Chicago: The University of Chicago Press, 1994], 258). Schafer admitted to taking advantage of his new construction's associations, "intending it to be a nervous word. Related to schizophrenia, I wanted it to convey the same sense of aberration and drama" (R. M. Schafer, *The Soundscape: Our Sonic Environment and the Tuning of the World*, Rochester, VT: Destiny Books, 1994 [1977], 91). In my use of the term here, I do not mean to imply an endorsement of Schafer's misgivings but instead hope to evoke its most literal meaning—that is, the split (*schizo-*) of sounds (*-phonia*) caused by the recording medium. Schafer introduced the term in *The New Soundscape: A Handbook for the Modern Music Teacher* (Toronto: Berandol, 1969).

4. See F. K. Engel, "The Introduction of the Magnetophon," in *Magnetic Recording: The First 100 Years*, ed. E. D. Daniel, C. D. Mee, and M. H. Clark (New York: IEEE Press, 1999), 47–48.

5. Emile Berliner introduced the disc, which would ultimately win out over Edison's cylinder. The disc's critical advantage was less its sound quality (though it was louder) than its practicality—it was easier to operate and, because of its flat shape, easier to store (Millard, *America on Record*, 125–35).

6. M. H. Clark, "The Magnetic Recording of Sound," in *Magnetic Recording: The First 100 Years*, ed. E. D. Daniel, C. D. Mee, and M. H. Clark (New York: IEEE Press, 1999), 7–8.

7. D. L. Morton Jr., *Sound Recording: The Life Story of a Technology* (Baltimore, MD: The Johns Hopkins University Press, 2004), 50–54, and M. H. Clark and Henry Nielsen, "The Telegraphone," in *Magnetic Recording: The First 100 Years*, 15–29.

8. A condenser microphone, which converted sounds into electric currents, replaced the recording horn, and those electronically amplified currents drove the movements of the cutter, rather than the diaphragm's vibrations (Millard, *America on Record*, 141).

9. By the term *phonograph* I'm referring to the technology of mechanical or electromechanical recording medium and not to the particular invention of Edison.

10. Millard, *America on Record*, 196, and D. Morton, *Off the Record: The Technology and Culture of Sound Recording in America* (New Brunswick, NJ: Rutgers University Press, 2000), 136–70.

11. Eduard Schüller, who worked for AEG, invented the "ring head" (patented in 1933), whose smooth surface was much better suited to the delicate tape than the previous chisel-shaped head, which often shredded the tape (Engel, "The Introduction of the Magnetophon," 51–52).

12. Ibid., 47–61. In 1935, AEG introduced the Magnetophon K1 at the Berlin radio exhibition, but critics promptly concluded that the sound quality remained inadequate and lamented the fact that the reproduced sound signal bore significant noise. The model K4 came three years later, after further experimentation and improvements that included replacing the carbonyl iron with magnetite. Magnetophon combines the words *magnetic phonograph*, while *K* stands for *Koffer*, which is the German word for "portable case" (Ibid., 54–56).

13. Ibid., 67.

14. B. R. Gooch, "Building on the Magnetophon," in *Magnetic Recording: The First 100 Years*, 73.

15. M. H. Clark, "Product Diversification," in *Magnetic Recording: The First 100 Years*, 92–93.

16. Gooch, "Building on the Magnetophon," 83–89.

17. Clark, "Product Diversification," 92–93.

18. R. H. Snyder, "Sel-Sync and the 'Octopus': How Came to Be the First Recorder to Minimize Successive Copying in Overdubs," *ARSC Journal* 32(2) (2003): 209–13.

19. Ibid., 210.

20. D. Laing, "A Voice without a Face: Popular Music and the Phonograph in the 1890s," *Popular Music* 10(1) (1991): 7.

21. J. Sterne, *The Audible Past: Cultural Origins of Sound Reproduction* (Durham, NC: Duke University Press, 2003), 264.

22. Edison as quoted in M. Chanan, *Repeated Takes: A Short History of Recording and Its Effects on Music* (London: Verso, 1995), 24–25.

23. M. Katz, *Capturing Sound: How Technology Has Changed Music* (Berkeley: University of California Press, 2004), 2.

24. J. Toynbee, *Making Popular Music: Musicians, Creativity and Institutions* (London: Arnold, 2000), 70–73.

25. Quoted in Sterne, *The Audible Past*, 292.

26. In his description of the early use of the electric condenser microphone, Toynbee echoes Sousa's image: "The microphone stayed resolutely outside the ensemble, and the performing system—small group, orchestra, solo performer—was consolidated as a *hermetic unit*" (Toynbee, *Making Popular Music*, 71, emphasis added). While recordings were still a legitimate documentation of a preexisting performance in terms of time and space, in live settings, the electronic amplification of sounds did impact the early era of schizophonia in one significant way: it challenged the human ability to locate sounds spatially—that is, to understand distant sounds as signs of faraway sources and intimate sounds as signs of nearby sources. While the vocalist stood up on the stage, his or her voice appeared to be next to the listener. Moreover, the vocalist might be standing at the right side of the stage while the voice came from an amplifier placed at the left side. Although electronic amplification divided sounds from their sources in live settings, it did not usher in a completely new era of schizophonia; electromechanical recording still only allowed for one-track recording and suffered its predecessor's severe limitations upon overdubbing. It was the invention of magnetic tape that would ultimately revolutionize our conception of music, and of musical recordings.

27. Stereophonic sound was introduced almost at the same time as multitrack recorders. For a discussion of the development of stereophonic recording, see Morton, *Off the Record*, 39–42, and P. Doyle, *Echo and Reverb: Fabricating Space in Popular Music Recording, 1900–1960* (Middletown, CT: Wesleyan University Press, 2005), 224–27.

28. T. Day, *A Century of Recorded Music: Listening to Musical History* (New Haven, CT: Yale University Press, 2000), 26.

29. E. Eisenberg, *The Recording Angel: Music, Records and Culture from Aristotle to Zappa* (New Haven, CT: Yale University Press, 2005), 122.

30. Ibid., 122–23.

31. This separates my "documentary" paradigm from what Toynbee calls the "documentary regime," which "wants to repress all that the technosphere implies (new opportunities for framing performance) by going back to the ideal context of concert hall or bar where musical interaction with the audience is, purportedly, immediate" (Toynbee, *Making Popular Music*, 70).

32. P. Arthur, "Jargons of Authenticity (Three American Moments)," in *Theorizing Documentary*, ed. M. Renov (New York: Routledge, 1993), 133.

33. For a discussion of how popular music is often packaged in terms of live performances, see T. Gracyk, *Rhythm and Noise: An Aesthetics of Rock* (Durham, NC: Duke University Press, 1996), 75–78.

34. Benjamin, "The Work of Art in the Age of Mechanical Reproduction," 220.

35. Eisenberg, *The Recording Angel*, 92.

36. M. Renov, "Toward a Poetics of Documentary," in *Theorizing Documentary*, ed. M. Renov (New York: Routledge, 1993), 33.

37. Eisenberg, *The Recording Angel*, 89.

38. Ibid., 90. I would say instead that these two ideals now coexist.

39. Katz, *Capturing Sound*, 41.

40. Chanan, *Repeated Takes*, 120.

41. See G. Gould, "The Prospects of Recording," in *Audio Culture: Readings in Modern Music*, ed. C. Cox and D. Warner (London: Continuum, 2004), 115–26.

42. For a discussion of transparent and opaque mediation, see R. Brøvig-Hanssen, "Opaque Mediation: The Cut-and-Paste Groove in DJ Food's 'Break,'" in *Musical Rhythm in the Age of Digital Reproduction*, ed. A. Danielsen (Farnham, UK: Ashgate, 2010), 159–75.

43. Other participants in the Project of Music for Magnetic Tape were Earle Brown, Morton Feldman, Christian Wolff, and David Tudor. *Williams Mix* was created by John Cage, Earle Brown, and David Tudor, with the technical and creative assistance of Louis and Bebe Barron (T. Holmes, *Electronic and Experimental Music* [New York: Routledge, 2002], 114–15).

44. Quoted in Holmes, *Electronic and Experimental Music*, 116–17.

45. T. D. Taylor, *Strange Sounds: Music, Technology and Culture* (New York: Routledge, 2001), 46.

46. Groupe de Recherces Musicale was the first established major electronic music studio (Holmes, *Electronic and Experimental Music*, 86).

47. Ibid., 135.

48. Ibid.

49. Ibid., 92.

50. Ibid., 125.

51. The members of Gentle Fire, which was founded in 1968, included Richard Bernas, Graham Hearn, Michael Robinson, Stuart Jones, Richard Orton, and Hugh Davies.

52. H. Davies, "Gentle Fire: An Early Approach to Live Electronic Music," *Leonardo Music Journal* (11) (2001): 53.

53. Doyle, *Echo and Reverb*, 181–83.

54. Toynbee, *Making Popular Music*, 86.

55. Quoted in M. Lewisohn, *The Complete Beatles Recording Sessions* (London: Hamlyn, 1988), 72.

56. Ibid., 96. Ken Townsend, one of the sound engineers for this recording, explained the process: "I . . . came up with a method whereby I fed a 50 cycle tone from the track of one machine then raised its voltage to drive the capstan motor of the second, thus running the two in sync" (Ibid.).

57. Holmes, *Electronic and Experimental Music*, 78.

58. Ibid., 77.

59. Ibid., 116–17.

60. A. Feenberg, "Subversive Rationalization: Technology, Power, and Democracy," in *Technology and the Politics of Knowledge*, ed. A. Feenberg and A. Hannay (Bloomington: Indiana University Press, 1995), 9.

Bibliography

Arthur, Paul. "Jargons of Authenticity (Three American Moments)." In *Theorizing Documentary*, ed. M. Renov, pp. 108–34. New York: Routledge, 1993.

Benjamin, Walter. "The Work of Art in the Age of Mechanical Reproduction." In *Illuminations: Essays and Reflections*, ed. H. Arendt, translated by Harry Zohn. New York: Schocken Books, 1968 (1936), 217–42.

———. "The Work of Art in the Age of Its Technological Reproducibility." In *The Work of Art in the Age of Its Technological Reproducibility and Other Writings in Media*, ed. Michael W. Jennings, Brigid Doherty, and Thomas Y. Levin, translated by Edmund Jephcott, Rodney Livingstone, Howard Eiland, and others. Cambridge, MA: The Belknap Press of Harvard University Press, 2008 (1989), 19–55.

Brøvig-Hanssen, Ragnhild. "Opaque Mediation: The Cut-and-Paste Groove in DJ Food's 'Break.'" In *Musical Rhythm in the Age of Digital Reproduction*, pp. 159–75, ed. A. Danielsen. Farnham, UK: Ashgate, 2010.

Chanan, Michael. *Repeated Takes: A Short History of Recording and Its Effects on Music*. London: Verso, 1995.

Clark, Mark H. "The Magnetic Recording of Sound." In *Magnetic Recording: The First 100 Years*, ed. E. D. Daniel, C. D. Mee, and M. H. Clark, pp. 6–14. New York: IEEE Press, 1999.

———. "Product Diversification." In *Magnetic Recording: The First 100 Years*, ed. E. D. Daniel, C. D. Mee, and M. H. Clark, pp. 92–109. New York: IEEE Press, 1999.

Clark, Mark H., and Henry Nielsen. "The Telegraphone." In *Magnetic Recording: The First 100 Years*, ed. E. D. Daniel, C. D. Mee. and M. H. Clark, pp. 15–29. New York: IEEE Press, 1999.

Davies, Hugh. "Gentle Fire: An Early Approach to Live Electronic Music." *Leonardo Music Journal* 11 (2001): 53–60.

Day, Timothy. *A Century of Recorded Music: Listening to Musical History*. New Haven, CT: Yale University Press, 2000.

Doyle, Peter. *Echo and Reverb: Fabricating Space in Popular Music Recording, 1900–1960*. Middletown, CT: Wesleyan University Press, 2005.

Eisenberg, Evan. *The Recording Angel: Music, Records and Culture from Aristotle to Zappa*. New Haven, CT: Yale University Press, 2005.

Engel, Friedrich K. "The Introduction of the Magnetophon." In *Magnetic Recording: The First 100 Years*, ed. E. D. Daniel, C. D. Mee, and M. H. Clark, pp. 47–71. New York: IEEE Press, 1999.

Feenberg, Andrew. "Subversive Rationalization: Technology, Power, and Democracy." In *Technology and the Politics of Knowledge*, ed. A. Feenberg and A. Hannay, pp. 3–22. Bloomington: Indiana University Press, 1995.

Feld, Steven. "From Schizophonia to Schismogenesis: On the Discourses and Commodification Practices of 'World Music' and 'World Beat.'" In *Music Grooves: Essays and Prologues*, ed. C. Keil and S. Feld, pp. 257–89. Chicago: University of Chicago Press, 1994.

Gooch, Beverly R. "Building on the Magnetophon." In *Magnetic Recording: The First 100 Years*, ed. E. D. Daniel, C. D. Mee, and M. H. Clark, pp. 72–91. New York: IEEE Press, 1999.

Gould, Glenn. "The Prospects of Recording." In *Audio Culture: Readings in Modern Music*, ed. C. Cox and D. Warner, pp. 115–26. New York: Continuum, 2004.

Gracyk, Theodore. *Rhythm and Noise: An Aesthetics of Rock*. Durham, NC: Duke University Press, 1996.

Holmes, Thom. *Electronic and Experimental Music*. New York: Routledge, 2002.

Katz, Mark. *Capturing Sound: How Technology Has Changed Music*. Berkeley: University of California Press, 2004.

Laing, Dave. "A Voice without a Face: Popular Music and the Phonograph in the 1890s." *Popular Music* 10(1) (1991): 1–9.
Lewisohn, Mark. *The Complete Beatles Recording Sessions*. London: Hamlyn, 1988.
Millard, Andre. *America on Record: A History of Recorded Sound*. Cambridge, UK: Cambridge University Press, 1995.
Morton, David. *Off the Record: The Technology and Culture of Sound Recording in America*. New Brunswick, NJ: Rutgers University Press, 2000.
———. *Sound Recording: The Life Story of a Technology*. Baltimore, MD: The Johns Hopkins University Press, 2004.
Renov, Michael. "Toward a Poetics of Documentary." In *Theorizing Documentary*, ed. M. Renov, pp. 12–36. New York: Routledge, 1993.
Schafer, R. Murray. *The New Soundscape: A Handbook for the Modern Music Teacher*. Toronto: Berandol, 1969.
———. *The Soundscape: Our Sonic Environment and the Tuning of the World.* Rochester, VT: Destiny Books, 1994 (1977).
Snyder, Ross H. "Sel-Sync and the 'Octopus': How Came to Be the First Recorder to Minimize Successive Copying in Overdubs." *ARSC Journal* 34(2) (Fall 2003): 209–13.
Sterne, Jonathan. *The Audible Past: Cultural Origins of Sound Reproduction*. Durham, NC: Duke University Press, 2003.
Taylor, Timothy D. *Strange Sounds: Music, Technology and Culture*. New York: Routledge, 2001.
Toynbee, Jason. *Making Popular Music: Musicians, Creativity and Institutions*. London: Arnold, 2000.

CHAPTER 6

Stockhausen Meets King Tubby's

The Transformation of the Stepped Filter into a Musical Instrument

Sean Williams

Introduction

This research was provoked by an interest in the similarity of particular sounds in the music of Karlheinz Stockhausen and Osbourne Ruddock (aka King Tubby) and an interest in the sources of these sounds and the musicianship that lies behind them. I am focusing on one particular type of technology used by both of these music makers and its use as an electronic musical instrument in these two very different contexts. The physical instruments are not identical, but they operate around a very similar principle, and above all, they *sound* very similar in both cases.

These sounds stem from the use of a stepped filter—in Stockhausen's case a band-pass filter; in Tubby's, a high-pass filter—and the sonic results owe a great deal to the repurposing of the original devices for use as performance tools, or musical instruments rather than static, technical devices. This chapter explores the relationships between technology design and practice and shows how information obtained via a material study of the technology can influence the critical study of both composition and performance practice.

On the face of it, there is no obvious link between Tubby's dub mixes such as "Rebel Dance"[1] and "Tubby's Dub Song"[2] and Stockhausen's *Mikrophonie I*[3] in terms of style, genre, audience, listening space, or any other conditions, and I neither make the case for any communication between King Tubby and Stockhausen, nor do I suggest that they were even aware of each other's work. However, if we leave aside the social, circumstantial, geographical, and economic factors and, through critical listening, concentrate on the sounds heard on these records and in these performances, we can begin to appreciate some fundamental sonic similarities and trace these back to similarities in the human musical performance.[4] To go one step further, I suggest that this particular music is not just created by straightforward musical performance but by a combination of musical imagination and technical expertise—an "alignment"[5] between practice and technical knowledge that challenges and repurposes existing technology, adapting it to use in creative music practice.

Filters

The tone control on a radio is the simplest example of an everyday filter. Fully open (set to 10), the filter allows the entire audio signal to pass through with no effect. As the tone control is turned down, the higher frequencies in the signal are attenuated while the lower frequencies pass through unaffected. This common tone control is a low-pass filter with a very gentle slope. Another common example of filtering is the experience of listening to a hi-fi from an adjacent room with the door closed. Most of the high frequencies will be absorbed by the door and wall, but the lower frequencies will still be audible. This is also a low-pass filtering effect but with a steeper slope and a lower cut-off frequency; that is, the frequency above which the amplitude is reduced.

High-pass or band-pass filters are less common as purely acoustic effects, but the telephone is a ubiquitous electronic example of a band-pass filter. Since the intelligibility of the human voice relies on a relatively narrow band of frequencies, it is possible to discard, or to filter out, both very low and very high frequencies without sacrificing intelligibility. Typically a telephone will not reproduce frequencies above 3.5 kHz (a low-pass filter) and not below 350 Hz (a high-pass filter). Since there is a low-pass and high-pass filter in series, we can consider this to be a band-pass filter; that is, all frequencies within the filter's frequency band are allowed to pass through while all frequencies outside the pass-band are attenuated.

Mikrophonie I and the Maihak W49 Filter

Designed in 1950 by Maihak AG of Hamburg, the W49 filter was first used by Stockhausen for the realization of *Kontakte*[6] between 1958 and 1959. The composition (and subsequent performances) of *Mikrophonie I* from 1964 onward marks the start of Stockhausen's regular use of a pair of these filters in many pieces, including *Kurzwellen*, *Prozession*, *Hymnen*, and *Aus den Sieben Tagen*, usually to filter the tam-tam and the viola. Starting in the 1950s with the use of electronic sound sources manipulated via the precise cutting and editing of tape for *Electronic Study I* and *II*, Stockhausen went on to create *Gesang der Jünglinge* in 1956, which boldly combined electronic sound sources with acoustic or *concrète* sounds. Subjecting sine-wave and pulse-wave generators as well as the acoustic sound of a child's voice to extreme tape editing, speed manipulation, electromechanical reverberation, and filtering, and presenting the piece as a fixed tape piece in four channels,[7] Stockhausen firmly established himself at the cutting edge of the new music, fully embracing the latest technological materials. He was based at the Westdeutscher Rundfunk (WDR) in Cologne where he was director of the Electronic Music Studio for a large part of his career and was able to draw on their huge technical resources while working there.

It should be recognized that the instruments used to create these early works were mostly laboratory test equipment, including oscillators, filters, and meters. In the mid-1950s there were only isolated instances of dedicated electronic musical instruments, but by 1964 when Stockhausen was writing *Mikrophonie I* the use of electronic instruments was far more widespread, with dedicated electronic music studios in many major cities and universities throughout Europe and the United States.[8] *Mikrophonie I* was another step forward in that it combined acoustic and electronic sound manipulation with live performance, but in the classical tradition, using a score and being presented in the concert hall. This piece evolved from Stockhausen's experimentation with a large tam-tam in his garden while his engineer, Jaap Spek, sat in the house and manipulated a filter and amplifier to change the sound picked up from a microphone, recording the result to tape. In a live performance of *Mikrophonie I* there are two groups of three performers. In each group, one excites the tam-tam by hitting, scraping, bowing, shouting at it, and other unconventional methods; one uses a microphone to pick up sounds at varying distances from the surface of the tam-tam and from the site of excitation, also using a resonator such as a cardboard tube, box, or wineglass to acoustically filter the sound; the other sits in the audience and controls a W49 filter and two volume faders, changing the timbre as well as the amplitude and position in space of the resulting sound. The loudspeakers are positioned at four corners of a square, giving the audience an immersive sound, but it should be remembered that the acoustic sounds are also audible to a greater or lesser degree. Rolf Gehlhaar, who performed the piece with the Stockhausen Ensemble between 1966 and 1970, describes the process:

> The tamtam is so strong that the loudspeaker sound has a bit of a struggle sometimes, and if it doesn't come out of the loudspeakers then the filters are useless. The beauty of the piece is exactly that polyphony between the amplified and the filtered—that's why the filters are important because you get an amplified sound which is different from the unamplified sound, so it's quadraphonic, or polyphonic.[9]

The score gives precise directions for each performer and also offers additional information about the choice of implements with which the tam-tam must be excited.[10] Great detail is given about the electronic instruments used for performing the piece, and this is where our examination of the filter, the Maihak W49 Hörspielverzerrer, begins (Figure 6.1).

Bandpass filter

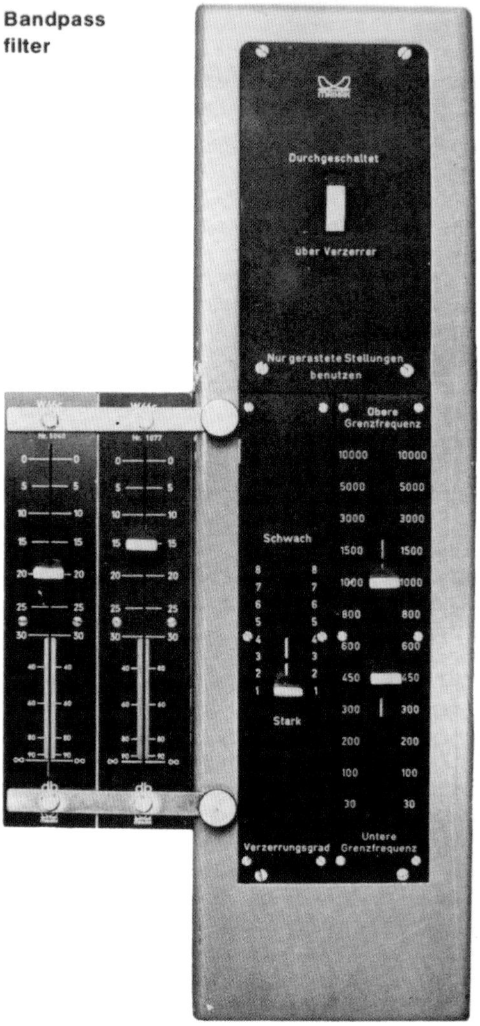

Filter W 49 „Hörspielverzerrer" ("Radio play distorter"), by the firm MAIHAK, Hamburg, Semperstrasse 38.

Two slide potentiometers are screwed on to the left side of the filter by means of 2 metal strips.

Two such filters, each with two potentiometers, are fastened to small tables 40 cm high in one row of seats in the middle of the hall, about 5 metres apart.

At best the adjustable 4-channel preamplifier for the loudspeakers is placed directly beside one of the two filterers.

Frequency scale

Setting of the attenuation slope always maximum (Verzerrungsgrad 1)

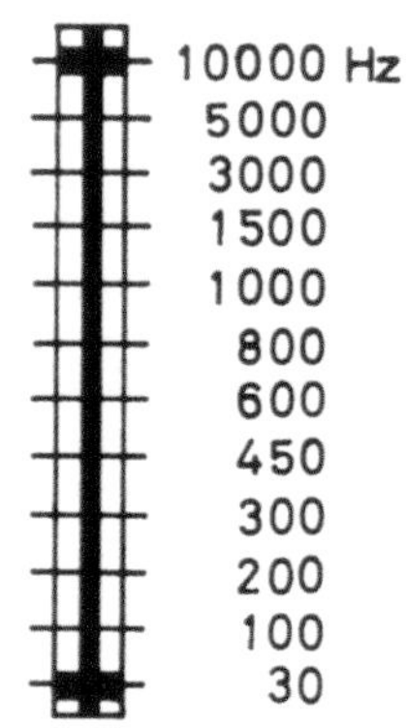

upper and lower cut-off knobs can be moved individually, each one as far as the other's position: the greatest bandwidth is 30–10,000 Hz, the narrowest corresponds to the distance between two adjacent frequency numbers. The frequency range *between* the two knobs is audible. The cut-off knobs can be moved by *one* hand in all positions.

FIGURE 6.1

W49 filter and two W66C faders. From Karlheinz Stockhausen "*Mikrophonie I,*" London: Universal Edition, 1974. Copyright © 1974 by Universal Edition (London) Ltd., London/UE15138. Courtesy of the Archive of the Stockhausen Foundation for Music, Kürten (www.stockhausen.org).

Thanks to the extensive documentation in the scores for both *Mikrophonie I*[11] and *Kontakte*,[12] we can get a reasonably good idea of how this instrument was used and what it was used for.

In Figure 6.1 we see the W49 filter with two Maihak W66C volume faders bolted to it, together comprising the electronic performance instrument used by each of the two performers in *Mikrophonie I*. Of the handful of W49s I have seen (Serial Nos. 6, 16, 21, 34, 69, 071, and 119) all except one (Serial No. 071) have been consistent with that pictured above and described in the technical documentation. Costing around 6,000 Deutsche Mark in 1969, only a few units were made, and these were only affordable by the large broadcast companies such as Westdeutscher Rundfunk. The unit weighs a hefty 10.9 kg (enough to incur excess baggage charges on certain budget airlines) and accepts a balanced input signal that passes through a low-pass and a high-pass filter to a balanced output. Filter slope and frequencies are adjusted in steps by means of three sliding switches. We will examine the sonic and performance characteristics of the W49 with direct comparison to the filter used under very different circumstances by King Tubby.

King Tubby's MCI Mixing Desk

King Tubby had set up his studio in a bedroom of his mother's house at 18 Dromilly Avenue, Kingston, Jamaica, originally working on a homemade mixing desk, and using the place as a workshop for building and maintaining his sound system, "King Tubby's Hometown Hi-fi." The bathroom was used occasionally to overdub vocals, but the studio was used almost exclusively for mixing four-track tapes that had been recorded at other studios on the island. The mixing desk that superseded the original homemade device was originally bought and probably commissioned by Byron Lee for his recently acquired Dynamic Sounds studio in Kingston, Jamaica, in 1969 for use with four-track tape-recording equipment. Lee was no stranger to new technology, having been one of the first to bring to Jamaica a Fender electric bass and amplifier in 1959 to allow his band The Dragonnaires greater flexibility when touring.[13] Given his connections and the close proximity of Jamaica to Miami, it is possible that he was influenced in his choice of mixing desk thanks to the great reputation and huge number of hits produced on the large MCI desk specially built by Grover C. "Jeep" Harned at MCI for Criteria Studios in Miami. It wasn't until 1972 that Harned developed the early, standardized design for the MCI JH400 series,[14] and until then, mixing desks were often custom designed and not readily available as generic production models. Although similar in some respects to a desk made for King Studios, I have seen no evidence that this particular desk was one of a series, and no information about this desk appears in any of the available MCI literature.

When Byron Lee upgraded Dynamic Sounds in 1972 from four-track to eight-track capability, Tubby was persuaded by the producer Bunny Lee (no relation to Byron) to purchase this mixer along with the old Ampex and Scully four-track tape machines, now considered obsolete by Byron Lee, to upgrade his own studio. Until purchasing this MCI desk, it seems that Tubby had been relying on a smaller homemade mixer, which had no capacity for four-track mixing.

Although limited by today's standards, the MCI desk offered twelve input channels, four output groups, remote tape transport control, a test tone oscillator, and an unusual high-pass filter built into the top right-hand corner of the master section (Figure 6.2).

One of King Tubby's engineers, Lloyd "King Jammy" James, recounts some details about the mixer:

> It was a very unique board because it was custom built for Dynamic Sounds . . . it had things that the modern boards nowadays don't really have, like a high-pass filter that made some squawky sounds when you change the frequency . . . We would put any instrument through it—drums, bass, riddim, voices. That high-pass filter is what create the unique sound at Tubby's.[15]

Reggae journalist Chris Lane refers to the filter as Tubby's "secret weapon,"[16] although he acknowledges that it was just one of many tools that contributed to Tubby's individual sound.[17]

FIGURE 6.2
King Tubby's MCI mixing desk. Courtesy of the Experience Music Project, Seattle.

Making a comparison of the filter frequency specifications of the MCI high-pass filter and the high-pass section of the W49 filter, it is possible to see the similarities in terms of the number of steps available and the relative frequency values in the two filters, both contributing to their similar sonic characteristics. Figure 6.3 shows a plot of the frequency steps in each filter.

The most obvious difference here is the greater definition in the low to mid range of frequencies of the W49. Between 100 Hz and 1 kHz, the W49 has seven steps, whereas the MCI filter has only five steps. This means that the W49 filter has better resolution throughout the midrange, where the MCI filter has slightly more control at the extreme ends of the frequency spectrum, and this is evident in many of Tubby's mixes, especially at the higher frequencies.

The values used in this chart were obtained from reading the legend on the MCI desk that now resides in the collection of the Experience Music Project in Seattle, after having been decommissioned from Tubby's studio some time after it stopped being used in the late 1980s and confirming these by examining other examples of the same model filter. The data for the W49 was gathered from Stockhausen's score, the examination of several W49 filters at the old Westdeutscher Rundfunk Studio in Cologne,[18] an undated technical document from Telefunken,[19] the original datasheets from Maihak AG,[20] and from a thorough examination of W49 serial number 071, which was kindly made available by the Musikinstrumente & Design Online Museum in Berlin.[21] It must be noted that it has not yet been possible to take any frequency measurements of these devices, so the data represented above remains unconfirmed beyond the manufacturers' specifications.

Repurposing

W49 Hörspielverzerrer

Before examining each instrument in depth, it is worth considering the original purposes of both of these devices in order to gain some insight into the musical approach behind the subsequent repurposing. It may seem like an obvious step to start using these devices in the way I am describing, but if it had been that obvious then there would surely be many more examples. By considering the repurposing as an appropriation and reinvention of the means of production made possible by the culture-technology alignment, it is possible to understand the significance of this in a broader cultural context. Neither Stockhausen nor King Tubby were prepared to make do with the standard uses of the available tools. Both used the transformation of the tools as an active musical practice, which in turn influenced the way they made music through feedback processes.

The original Braunbuch datasheet describes the W49 as a "Verzerrer für Hörspielzwecke"; that is, a distorter for radio-play purposes. According to this document it was designed "for the creation of acoustic effects by electronic means using frequency cutting."[22] Telefunken also produced a short, undated document in German, French, and

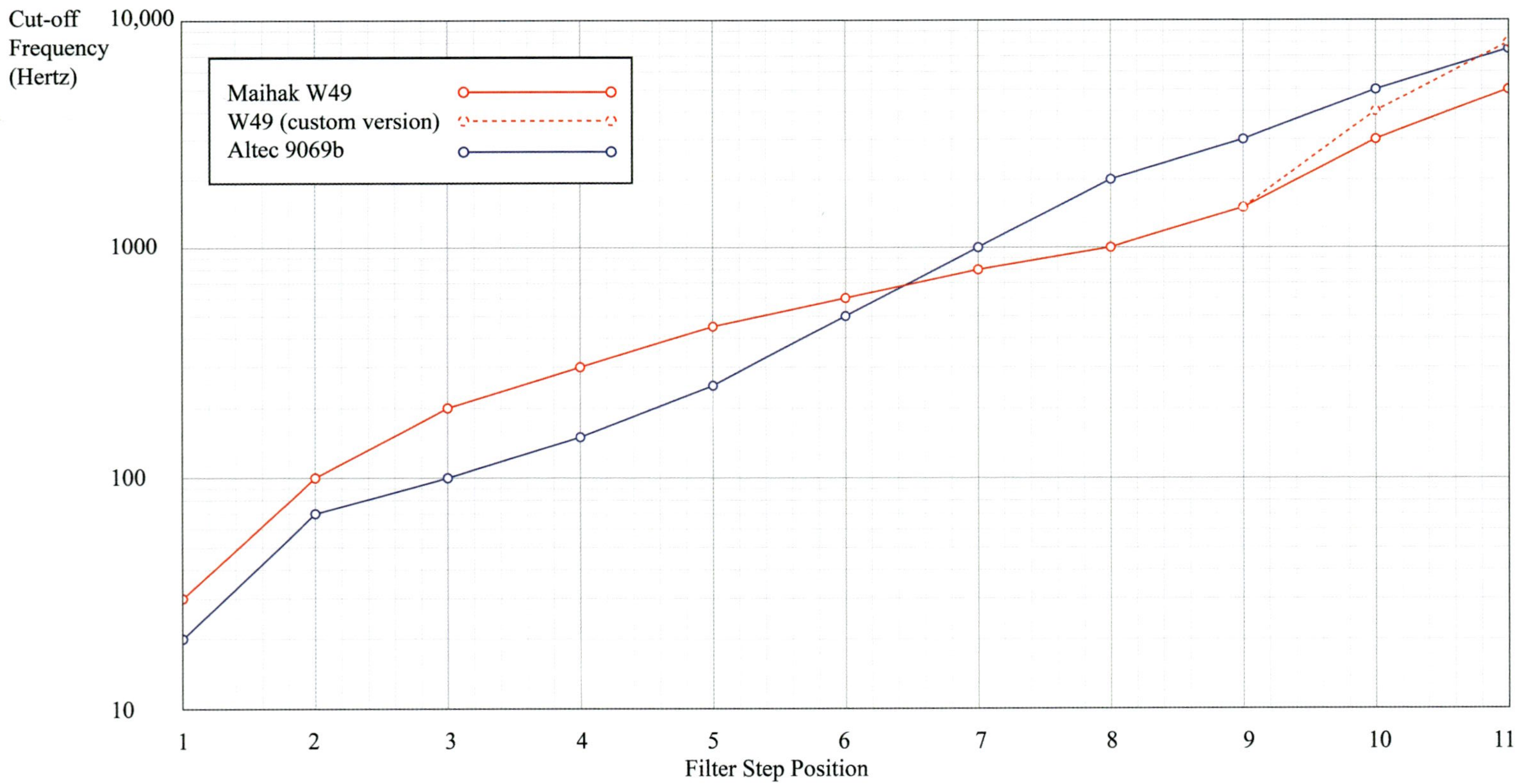

FIGURE 6.3
Comparison of filter frequencies specified for the W49 and the MCI/Altec 9069B (Sean Williams).

English detailing the W49.[23] The German name *Hörspielverzerrer* translates as *radio play distorter*. The French name *Correcteur de son pour pièce radiophonique* translates as *equalizer for radiophonic productions*. In this technical document, the English name is simply *Attenuating Equalizer*, leaving no implicit cue as to its intended usage; however, it is clear that it was designed for transforming sound in the context of radiophonic sound design, such as simulating distance, telephone conversations, and more.

This usage is still rather open-ended, but printed clearly above the frequency switches and unmissable to any user is the following instruction: "Nür gerastete Stellungen benutzen"; that is, "only used detented/notched settings." This is where Stockhausen's will clashes with the original purpose of the design since his score for *Mikrophonie I* instructs the filter performers to continually adjust the filter frequencies in direct contravention of the legend printed on the instruments.[24] The consequences of this transgression will be revealed below, but we should also consider this repurposing in the context of the other performers in the piece. *Mikrophonie I* also involves two performers scraping, bowing, shouting into, and doing all sorts of other things to a tam-tam that diverge from the expected way of using such an instrument, and two other performers using microphones in a similarly unorthodox manner. The repurposing of the W49 filter is therefore consistent with the approach demonstrated throughout the piece in which Stockhausen is pushing each performer to extend the music-making potential of each instrument or tool in order to create a larger, polyphonic sound that goes beyond any of the individual elements comprising it.

MCI High-Pass Filter

On many mixing desks it is standard to have a high-pass filter on the input to each channel to eliminate low-frequency rumble. This is usually either on or off and fixed at one frequency, although sometimes this frequency is switchable, usually somewhere between 60 Hz and 150 Hz. Tubby's MCI desk has no filter on any channel, but it instead has one filter that can be patched into any part of the signal chain as needed. Uniquely, it has an enormous range from 70 Hz to 7.5 kHz in ten steps, plus an "off" position. This extreme range questions its use as simply a rumble filter—7 kHz is almost an octave above the highest note on a grand piano—but it is not inconceivable that such high-frequency settings could be useful to enhance the sound of hi-hats or other high frequency instruments. Without reference to any of Harned's documentation it is difficult to know exactly what he envisioned for this feature; however, on close examination of the desk itself it appears that many components, such as the faders and possibly the EQ sections, are modular items. The filter itself is an Altec 9069B. This has identical specifications as the Langevin EQ255A, which cost $125 in 1968 and is further associated with the designer Art Davis, who was a common link between all of these companies that were variously taken over by and merged with each other around this time. Many mixing desks designed in the United States at this time used stock items such as faders, EQ sections, and VU-meters made by these companies, and King Tubby's MCI mixing desk is a good example of this design philosophy. The 9069B high-pass filter and its partner, the 9068B low-pass filter, were also sold combined into one rack-mount unit, the 9067B, which was marketed primarily as a sound-effects filter to the motion picture industry. Altec was also a key manufacturer of sound amplification equipment and had extensive experience with filter design, especially for crossovers in speaker systems. It is a neat coincidence that this filter with such a sound-system pedigree ended up in the studio of one of the world's most renowned sound-system pioneers.

Like the W49, this filter would typically be set and left alone, but, like Stockhausen, and unlike Syd Bucknor and other engineers at Dynamic Sounds, Tubby could not resist dynamically changing the filter's frequency, and through this active performance, treating it like any other musical instrument. The evidence is clearly audible in both his own mixes and those of his assistants/apprentices, including Philip Smart, Lloyd "Prince Jammy" James, and Overton "Scientist" Brown.

As in the case of *Mikrophonie I*, the repurposed filter was only one repurposed device among many, making it less surprising that it was used in this way. Tubby used one of the four-track tape machines fed from one or two group outputs as a tape delay/echo,[25] with an elegant use of a channel fader as both delay and feedback level, allowing for simplified use and ease of performance while mixing. He also used a customized Fisher K-10 Spacexpander—a domestic hi-fi reverberation device—as his main studio reverb, occasionally knocking or dropping the spring reverb tank to thunderous effect. A particularly good example is Tubby's dub of John Holt's "A Quiet Place," titled "A Noisy Place" and featured on *King Tubby's in Fine Style.*[26]

How Were the Instruments Used?

The evidence for *Mikrophonie I* has been gathered from film footage of an early performance, the score, and two interviews I conducted with Rolf Gehlhaar, who was Stockhausen's assistant and performed the piece many times as part of the Stockhausen Ensemble between 1966 and 1970. As shown in Figure 6.1, the W49 had two Maihak W66C volume faders bolted to the left-hand side of it, making it an integrated performance instrument. Gehlhaar came up with the solution of connecting the three units together with brass strips on joining the ensemble in 1966. Before this the W66C faders would move around too easily, but once attached to the heavy W49 filter, the setup became much more solid and so playable that it continued to be used in this configuration for another ten years.[27] The score demands that this whole apparatus is fastened to a small table 40 cm high,[28] and the height difference between the front and rear of the W49 ensures that, like the MCI mixing desk, the instrument ergonomically slopes slightly toward the performer. The combined weight of about 12 kg ensures that there would be little movement of the instrument during all but the most animated performance, and the formal concert hall setting with the performers seated in among the audience would also have constrained the performers from moving around too much during performances.

The evidence for Tubby's use of the MCI filter has been gathered from some film footage of Prince Jammy performing a dub mix with the desk (although sadly not using the filter),[29] accounts from Bunny Lee who produced many records with Tubby, and an interview I conducted with Chris Lane, who visited the studio in 1977[30] and there observed Prince Jammy mixing some dub versions. In addition I have built my own version of Tubby's filter, with particular attention paid to the control interface, and I have experimented at length to try to re-create some of the sounds and effects that he was able to produce. Subsequently, after identifying the exact filter used, I found an original unit and have since incorporated it into my own live performance setup. As a result of the parallels drawn up by this research, this has even been successfully used in a performance of Stockhausen's *Spiral* at the Soundings Festival in Edinburgh, 2011. Perhaps the most striking evidence of usage lies in the photos of the desk itself, which shows extensive wear patterns around particular controls. Critical listening to the many recordings made in Tubby's studio provides the main source of information (Figure 6.4).

Comparing the video clip of Jammy mixing "Jailhouse Rock"[31] with these wear patterns yields all sorts of additional information about performance practice, such as which channels were used for tape returns and delay returns, which channels were most often sent to the reverb unit, and what is most relevant here, how much use the high-pass filter frequency control was subjected to. The visual evidence for the latter is overwhelming, the ghost handprint with palm, thumb-base, and three knuckles clearly visible around the control (see Figure 6.4).

One or more tracks would be routed to the filter via a group output, with the resulting filter output being routed back to the main mix. This routing is mostly done in parallel with the original signal, thereby resulting in a mix between the original sound and the filtered sound. This mirrors the presence in performances of *Mikrophonie I* of the original acoustic sound alongside the amplified filtered sound both in live performance and in the recordings. The use of the filter alongside the fader control of different signal levels within a larger process is also a clear parallel.

This account by Bunny Lee reproduced on the back of the record sleeve of *Dub Gone Crazy* supports this hypothesis:

> An' Tubby's studio did 'ave a ting weh you could a thin it, an' do all different kinda ting with it, right,—it's not even really equalisation, the ting 'ave four push-up ting, when you push the one in the middle and 'ave it up and down, with the ting, it create some mad sound, like you hear all some knife a cut thru.[32]

Lee's description gives a good idea of what the filter sounds like. When filtered, the sound seems thinner, and the effect is easily likened to something being cut with a knife. The "four push-up ting" that Lee refers to are the four group faders, colored red, blue,

FIGURE 6.4
Detail of wear patterns around the filter control of the MCI mixing desk. Courtesy of the Experience Music Project, Seattle.

green, and white, with group 2 or 3 ("the one in the middle") being used as the high-pass filter send. To "'ave it up and down, with the ting" seems to refer to changing the filter frequency using the knob.

Performance

The most obvious factor about the MCI filter performance interface, apart from the ¾" diameter skirted knob, is the 15° angle between switch positions that allows all settings to be reached within 165° (just under half a revolution). In performance terms this influences the extent of filtering that it is possible to do in conjunction with additional manipulation of other parameters during a live mix. Indeed, the filter is often referred to as the "Big Knob"[33] filter, but the significance of a big knob is the increased precision available to the operator. Notice that only the master volume and talkback knobs are larger than the filter knob, with all others being of a smaller size.[34] This easily selectable range can be compared to the W49 in that despite the slider design of the latter, all filter frequencies can be selected using one (albeit large) hand, leaving the other free to adjust different parameters—in the case of *Mikrophonie I*, the two volume faders.

This convenient control for accessing all possible settings can only have encouraged the use of this feature as a performance device, making live sound manipulation an easy option. In both cases the filter frequency is adjusted with the right hand, leaving the left hand free to manipulate volume faders and other controls. Indeed, it is striking that Tubby's and Stockhausen's repurposing of their respective filters sprang from improvisation in both cases. One of the main features of instrumental improvisation is a tendency to use extended techniques, coaxing unexpected or nonstandard sounds from existing musical instruments.

Bunny Lee comments on Tubby's approach to mixing:

> He do it all live, too. He don't build it up bit by bit, him just leggo' the tape and do his thing. You watch him, it like watching a conductor or a maestro at work. And of course every time it would be different. He always want to surprise people—I think he even want to surprise himself sometimes—and if he mix the same tune a dozen times you will have twelve different version.[35]

In this context it is easy to understand the value of an economy of performance controls with both delay level and feedback being controlled by one fader, and the summed signal from multiple tape return channels being routed via one fader to the high-pass filter rather than individually. This arrangement allows for great control using only a few faders, thus enabling Tubby to construct and perform very complex mixes. Where the original purpose of the four output groups would have been to route microphone signals to the four channels of a tape machine, Tubby used them to increase the performative scope of the mixing desk—his instrument. Mixing to mono, he had three spare output groups and was able to use these for echo (delay) and filtering instead.

In order to make a comparison between Stockhausen's use of the W49 and Tubby's use of the MCI, I have transcribed a mix of "Rebel Dance,"[36] which features use of reverberation and the high-pass filter. There is no suggestion that Tubby worked from a score, but the intention is to compare the performance on the stepped filter to that indicated in the *Mikrophonie I* score and ultimately to use a similar filter to explore further the performance practice involved. Figure 6.5 is a detail of a page from *Mikrophonie I.*

This notation is for three performers; the tam-tam player follows the top line; the microphonist follows the middle section and is directed to hold the microphone near or

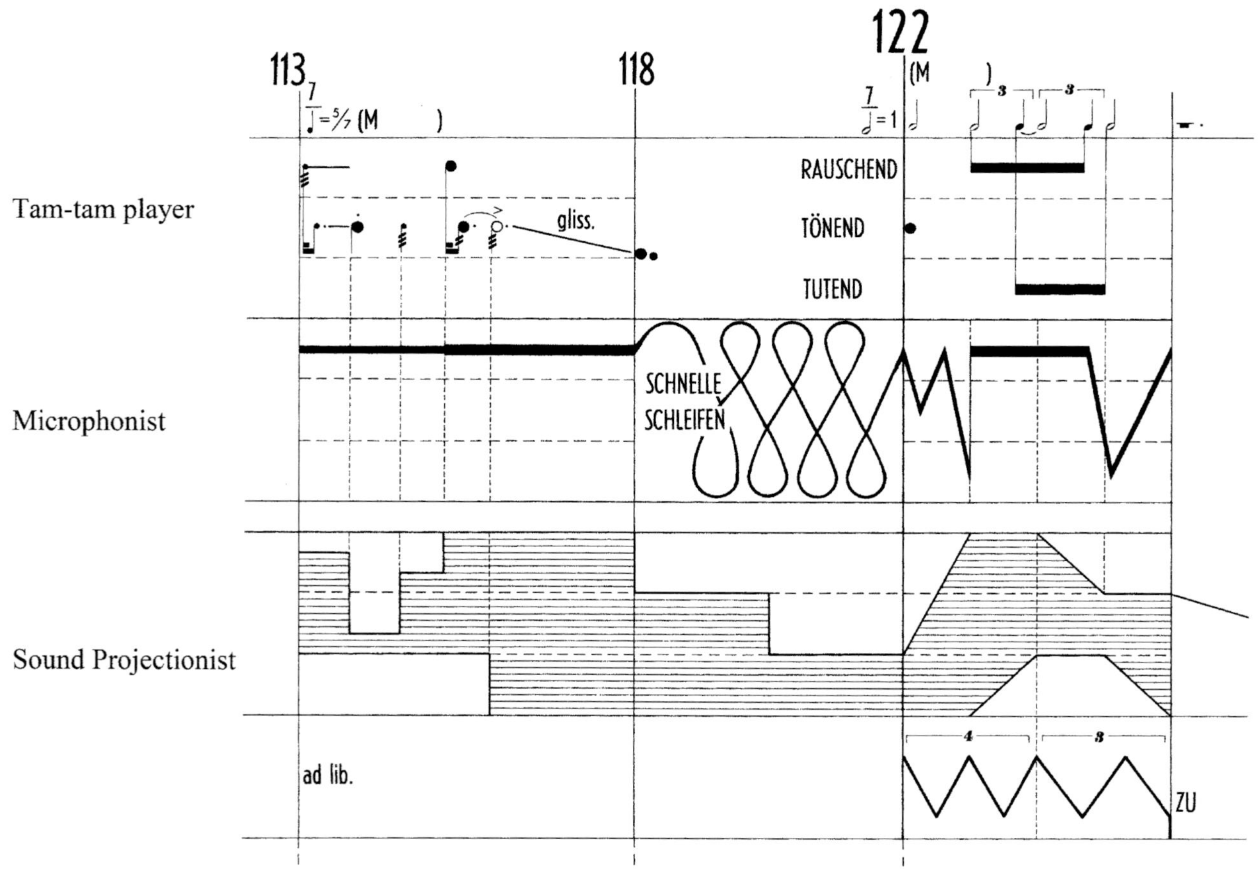

FIGURE 6.5
Detail from a page of the score for *Mikrophonie I*. From Karlheinz Stockhausen, "Mikrophonie I," © 1974 by Universal Edition (London) Ltd., London/UE15138.

far from the surface of the tam-tam and also near or far from the site of excitation; the sound projectionist follows the lower section, the shaded area representing the frequencies allowed to pass through the band-pass filter, and the line in the bottom section represents overall amplitude as governed by the W66C potentiometers attached to the left-hand side of the W49 filter (see Figure 6.1).

Concentrating on the filter notation, the two features most notable in this excerpt are the clear steps from measure 113 to measure 122, and the notated slopes from measure 122 to the end. This points to a similar usage of the filter as Tubby's for both sweeps and rhythmic punctuations, again using the stepped nature of the filter for rhythmic musical effect. The tie lines between the filter steps near the bottom of the page and the tam-tam events on the top row are testament to this, the filter frequency steps being coordinated with strikes of the tam-tam.

In the transcription of "Rebel Dance" shown in Figure 6.6, I have deliberately emulated Stockhausen's sound projectionist notation. Instead of transcribing the notes played by drums, bass, guitar, organ, and horns I have simply indicated the amount of each instrument in the mix as governed by the volume faders on the mixing desk. Since the mixing was done from four-track tape, guitar and organ were grouped together on one track. This also reflects the fact that no instruments were present at the time of mixing—that part of the performance in effect being done by the tape machine. Drums are present throughout, with the bass guitar only dropping out between measure 12 and 13 while the other instruments are used only sparingly. With measure numbers being read from left to right, the top row shows the filter frequency with the shaded area representing the frequencies allowed to pass through the filter. Note the lack of an upper frequency limit due to this filter not having a low-pass element. The second row shows the amplitude of the filtered signal just like the *Mikrophonie I* score, and the next four rows show the amplitude of each of the four tape tracks. Reverberation is applied individually to each tape return, but notation for this has not been included here.

At measure 16 there is a clear use of the discrete filter frequency steps to create a triplet punctuation, whereas in many other parts of the mix the filter is perceived as sweeping in a more linear fashion. This is an illustration of the steps in the filter and the related transients being used to impart additional rhythmic elements to the underlying sound, just as is evident in many places in Stockhausen's score. Indeed, Stockhausen uses stepped notation as well as slopes to imply this very distinction. In the score, he writes: "All filter changes notated as graphically continuous actually occur stepwise, due to the *stepped* filter."[37] Stockhausen, of course, specifies the usage of the filter in the score, but by his own admission, the score is the formalized result of a series of experiments he performed with the tam-tam in his garden and the filter in his kitchen.

What comes across from the performers of *Mikrophonie I* is an initial frustration with the audible clicks and the physical resistance of the faders. Gehlhaar recounts how they had to use contact cleaner to get rid of the frustrating clicks, and also how their fingers were painful and tired after a performance. He did not speak about the filter

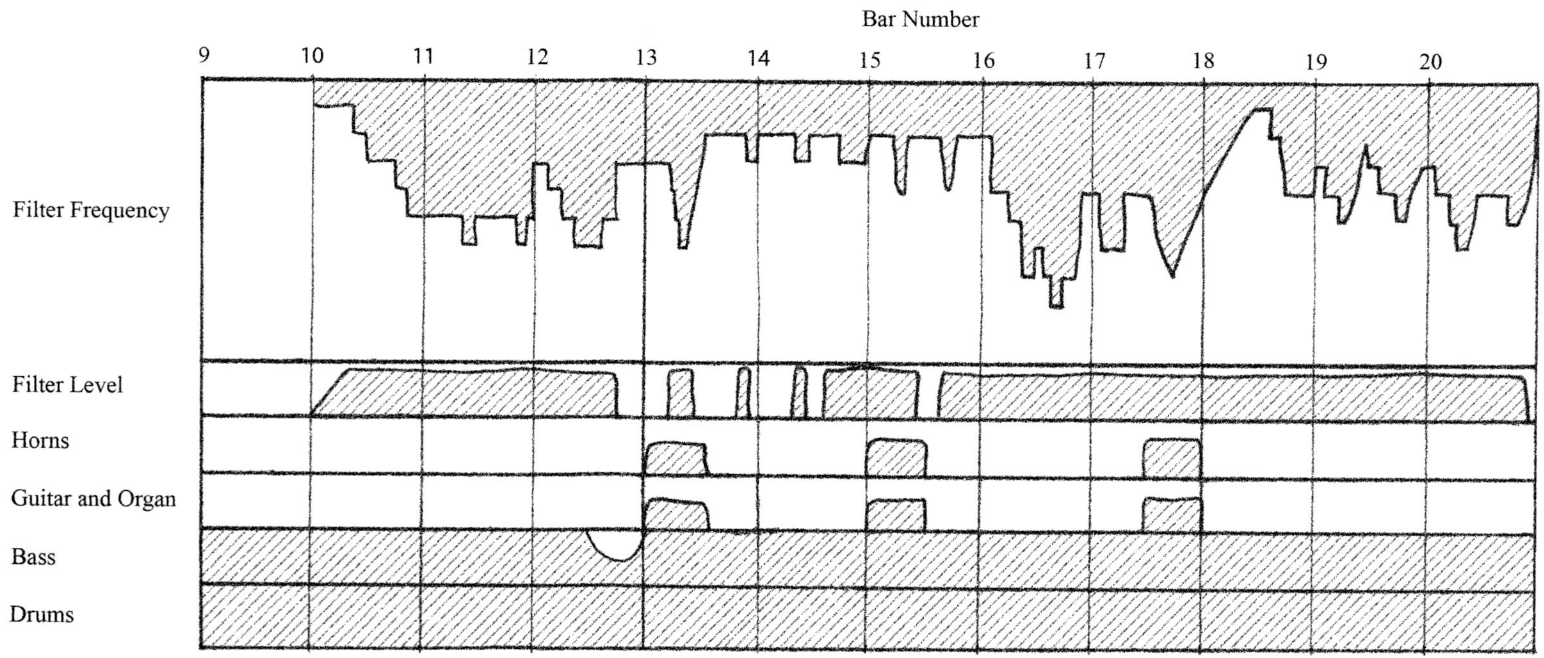

FIGURE 6.6
Transcription of part of "Rebel Dance" (Sean Williams).

fondly.[38] The departure of the performance from the "click-free" demands of the score is subject to Stockhausen's postrationalization in a lecture delivered in 1991:

> They were so-called Hörspiel-Verzerrer W49, built in-house at the WDR in Cologne: filters with carbon strips. It is really interesting how very old-fashioned that sounds (after all, violins with catgut are used today). Such materials are glorious, aren't they? The two metal levers of the filters scrape along on the carbon strips, and spray must now and then be used . . . Today, if you try to substitute computerized filter simulations, the characteristic sound goes to hell. The scraping and the skips between filter-levels is lost; but they actually belong to such a sound, when it is brightened up from below to above, or vice-versa. The score is also written in such a way that both controls can be opened and closed in the span between the index-finger and the little-finger of a spread hand. The W49 filter was quite fumblingly designed.[39]

This resonates with Jammy's comments about the MCI filter making "some squawky sounds when you change the frequency."[40] Used in this way, the W49 was never going to be a click-free filter, as we shall see, and it makes it all the more important for this to be taken into consideration by subsequent interpreters. It is also perhaps a little unfair to call the W49 "fumblingly designed" since it is quite clear from the legend printed on the unit itself that it was never meant to be used in this manner. The use of contact cleaner actually betrays a lack of technical knowledge. In a conversation with retired WDR Studio engineer Volker Müller, he told me about the musicians using contact spray to try to clean the switching mechanisms and how this inevitably made the clicks worse because it encouraged the buildup of dirt and debris.

What this all points to is the extent of influence over the final sound of the consequences of repurposing these instruments and the audible results of this process. This can be thought of as a translation of the physical presence of the device into an audible signature, but it can also act via its imperfections as a means by which the physical presence of the performer is revealed, and in this way it reinforces the individuality of both the instrument and the instrumentalist.

A useful mechanism by which we can analyze this is noise, both in the sense of unpitched or complex sounds not necessarily related to the input signal, and in the sense of the degradation of the signal between the ideal situation and the real-world phenomenon.

Stockhausen's Clicks

In the absence of any circuit diagrams we need to look inside the W49 to try to understand how the sound is filtered, but more importantly why it makes "clicks, crackles and pops," and how its construction and repurposing influences its use as a musical instrument (Figure 6.7).

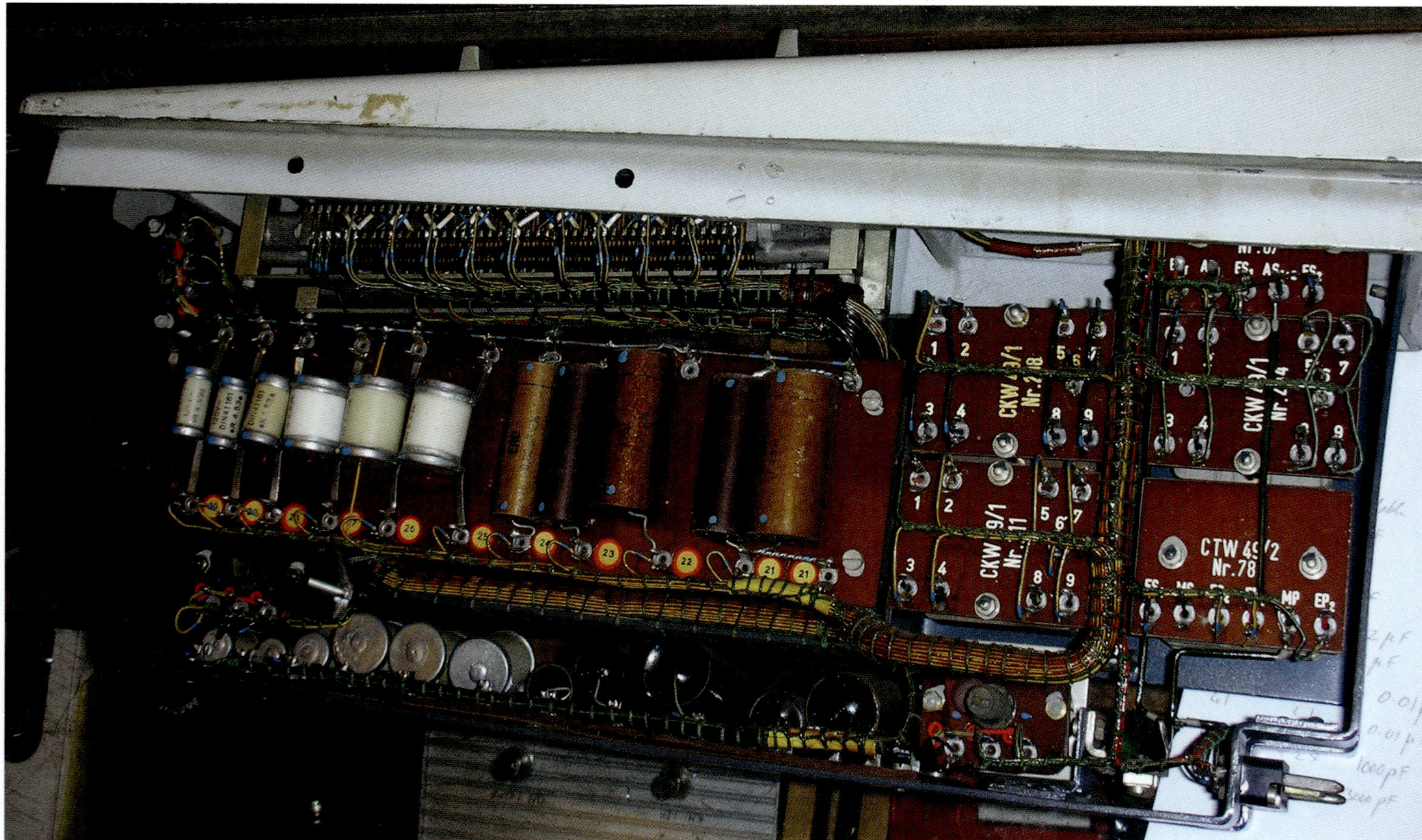

FIGURE 6.7
Internal detail of W49 serial 071. Courtesy Musikinstrumente & Design Online Museum, Berlin.

This is a passive filter based on an LCR network (inductors, capacitors, and resistors). To change the filter frequency a number of passive components are switched in and out of the signal path. No matter how much contact cleaner is used, this design will always click, pop, crunch, and crackle due to capacitors charging and discharging while being switched. The "click-free" condition specified by Stockhausen[41] was therefore never possible with this device. No amount of contact cleaner will fix this problem, although it is possible that dirty contacts could add to the switching noise.[42] This means that it is an inherent quality of the instrument, not in its original state but as a repurposed instrument, and indeed made worse by this repurposed usage. Considering the extent to which an effort was made by the manufacturer to advise against this usage by printing specific instructions above the controls, it should come as no surprise that the instrument exhibits its noisy behavior when used in contravention to this advice.

If we look at the components as shown in Figure 6.7, we see tolerances of between 2 percent and 20 percent. This implies that each individual W49 is likely to exhibit slightly different frequency responses and therefore sound different, and that the actual performance is likely to differ from the specified performance due to these deviations. Neither manufacturers' nor composers' specifications are necessarily realized accurately.

By measuring the force needed to move the faders—10 N—and the distance between the faders at their maximum span—160 mm—it becomes evident that some parts of the score may be impossible to perform unless the performers have large, strong hands. This is no surprise with Stockhausen's scores,[43] but it is worth pointing out in the context of the influence of the instrument on the performance. At the start of the following example, which is a detail from another page of the *Mikrophonie I* score (Figure 6.8), the faders must be moved from their maximum distance apart to very close together by the right hand while the left hand moves the volume faders in fast, small movements.

Gehlhaar's account of performances[44] is particularly enlightening in relation to the affordance of this instrument. He reports difficulties in seeing the onstage performers and following exactly where they were according to the score. This forced him to improvise, and he was afraid of Stockhausen noticing this until after a few performances he noticed that Stockhausen was having similar problems himself and was also having to improvise. Recall Bunny Lee's comment about Tubby's multiple mixes of the same song: "And of course, every time it would be different."[45] But a Tubby dub is still a Tubby dub, and a performance of *Mikrophonie I* by the Stockhausen Ensemble also retains an individual character, so there seems to be an important influence on the fidelity of the outcome exerted not only by the performers but also by the instrument itself and its imperfections, as alluded to by Stockhausen in his 1991 lecture.[46]

After a performance of the piece by the Anthos Ensemble given in Kürten in 2010 using a different filter without steps or clicks, Kathinka Pasveer noted that the sound was perhaps too clean and that Stockhausen preferred it when the filters were "kaput." This reinforces the idea that these imperfections eventually became vital to the sound of the instrument in its repurposed form and that the repurposing of the technology has an

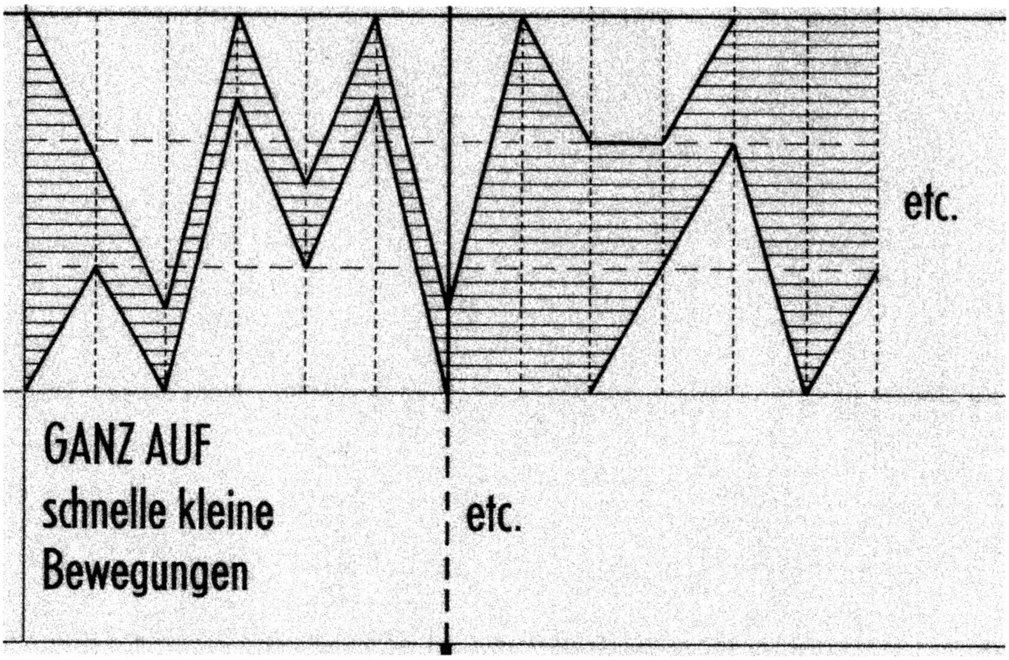

FIGURE 6.8
Detail from a page of the score for *Mikrophonie I*. From Karlheinz Stockhausen, "Mikrophonie I." Copyright © 1974 by Universal Edition (London) Ltd., London/UE15138.

additional impact on the composition and performance practice. Taken in conjunction with other directions, in the *Mikrophonie I* score to "ad lib" the physical nature of the filter and the way in which performers necessarily interact with it become vital to the successful/faithful communication of *Mikrophonie I* to an audience.

Tubby's Noise

We have noted that the clicks and pops from Tubby's filter are sometimes used as rhythmic punctuations and additions, but noise features far more in the dub mixes. The signal fed to the filter is complete with tape hiss, track bleed, reverb amp hum, and sympathetic vibration of the spring, among other noise. This all means that when the filter frequency is changed, the effect is far more noticeable, since all this background noise is also subject to filtering in addition to the original signal sent to the filter.

The Altec 9069B is based around a T-network of two capacitors and one inductor. To switch frequencies the inductor tap and both capacitors are switched by means of a three-pole rotary switch. There is little or no chance of click-free switching. The filter sounds fairly steep, and with the increased steepness of the filter slope comes more of a pronounced phase difference around the cut-off frequency. In the MCI desk, the filter was almost certainly intended to be used inline to remove low-frequency content from a signal recorded to tape; however, Tubby's desk used a filtered signal *in parallel* with an original signal, thereby making any such phase differences obviously audible due to phase cancellation. Since the background noise is spread over a wide frequency range, this phasing effect can sometimes be very pronounced and is audible at all filter settings, leading, as Chris Lane suggests, to the filter sometimes being called a phaser.[47]

In this way, the wider noise present in the recording process and the mixing process becomes a constituent part of the sound of the MCI filter used in this way by King Tubby and his assistants. It is the way it is used, beyond its original scope, that makes it sound this way, but that is also responsible for drawing attention to these sonic traces of physical interaction and the means of production. The inclusion of and interaction with the wider sound environment afforded by the use of the MCI filter directly affects the composition or sonic output, and this effect is what Stockhausen attempts to confront in his reflections on the characteristics of the W49 filter[48] through having had to deal with an imperfect and noisy instrument for many years and having learned to love it despite its flaws.

Conclusion

Although this research stemmed from a perception of sonic similarities in the work of King Tubby and Karlheinz Stockhausen, it has been possible to trace certain similarities at the heart of both of their approaches to making dynamic musical instruments out of otherwise static technical devices. The striking factors are the way in which noises and environmental sounds, often undesirable from a compositional point of view, have become part of the authentic sound despite the initial struggle against them. These imperfections end up by emphasizing the physicality of the performance and therefore the material nature of the music-making process, and so allow us, as listeners, a closer appreciation of the physical nature, of the *performance* by human beings of this recorded and partly electronic music. This performance aspect, which lies at the root of both of these very different music makers' approaches, is a vital part of what links their music-making activities in the absence of any other connection between them. It reaches across geographical, social, commercial, and economic boundaries, which otherwise position their work in diametric opposition to one another, and it sheds light on a common musical purpose through the joy of the musical manipulation of sounds using any available tools. The parallels of practice with these similar devices are not necessarily signs of technological determinism. They were not used in this way by other musicians, so in the cases of Stockhausen and King Tubby this practice is more a sign of a common approach and attitude toward shaping the world with the available tools, but by dialogue and feedback, both shaping and being shaped by the tools themselves.

It should, by now, be obvious that these kinds of electronic devices, whether used within their original operational parameters or extended and repurposed in the ways detailed above, must be considered as individual musical instruments and not simply as manufactured electronic tools. Their use as instruments relies on the relationship between their physical and electronic design, and bound up with this is the interface that they offer the performer and the relationship between the physical and aural feedback the performer experiences. But it is only in the context of the alignment between expertise and technology that they become musical instruments. The social context is all important.

By examining the mechanisms by which these two instruments have shaped the music, and by observing the physical interaction between performer and instrument, we can understand how a particular technology can have a two-way relationship with the composition, both shaping it and being shaped by it. Beyond that, a material study enables us to reevaluate such instruments and, with the benefit of hindsight, to discover certain characteristics, like Stockhausen's subsequent appreciation of the "scraping and the skips between filter-levels,"[49] which may have been struggled with at the time but have since been recognized as essential to the nature of the instruments and the music made with them. It is so often the case that a struggle to overcome certain limitations, here exemplified by the repurposing of both of these devices, yields far more interesting results than a situation in which everything is too easy.

As a musical instrument, the stepped filter only really features in repurposed form. Often used in much more limited frequency ranges in mastering studios or on some mixing desks, I am unaware of other physical examples exhibiting such a wide range of cut-off frequencies as these two units examined above. Improvements in electronic component specifications, voltage control systems, and eventually digital control eliminated the need for limiting filters used for music to discrete steps, so subsequent designs used in synthesizers featured continuous control instead. All the limitations, idiosyncrasies, and affordances of these stepped filters disappeared with the adoption of the new technology.

Given that it is possible to study a select few instruments such as these in museum collections as different in scope as the Experience Music Project and the Musikinstrumente & Design Online Museum, as well as being able to listen to recordings of these instruments in use, this kind of material research can provide a solid phenomenological foundation for further musicological, sociological, or anthropological studies of the music made using them. There are few better examples of this than Kehew and Ryan's detailed explanation of the tools and processes behind the original stereo mixes of the Beatles' albums[50] and the amount of speculation, theories, and imaginative stories that have been debunked since its publication. They give a detailed explanation of the recording process and explain why particular groups of instruments ended up on each of the four tracks of tape. Accurately describing the limited panning controls available on channels 1, 2, 7, and 8 of the EMI REDD.51 mixing desk (the channels used as tape returns), they show by way of a material study that the stereo mixes executed using this desk in Abbey Road's Studio 2 sound the way they do largely due to the affordance and limitations of the REDD.51 mixing desk.[51] Indeed, the desk was designed for two-track stereo recording of classical music with the possibility of adding up to four soloist microphones, so using it for discrete four-track recording was another case of repurposing.

For me, the most interesting material is the sonic evidence of music makers' physical interaction with the electronic sound-making equipment, and the noisy traces that locate this physical relationship in the wider ecological framework that extends beyond the composition through the reproduction media and the listening environment into the ear and the consciousness of the listener, brilliantly made visible by the wear patterns on King Tubby's MCI desk. The imperfections resulting from the struggle to make machines serve a purpose exceeding their original design limitations is a direct indication of the presence of human will and action in the creative music-making process and can be thought of as evidence of resistance to technological determinism. This human connection, extending all the way back past the performer to the design, engineering, and building of the instruments themselves, in this case by the staff at Maihak AG, Grover C. "Jeep" Harned at MCI, and probably Art Davis at Cinema Engineering and Altec, is a vital and dynamic part of what constitutes this electronic music.

Notes

1. Winston "Niney the Observer" Holness, "Rebel Dance," in *King Tubby's Special 1973–1976*, ed. Osbourne Ruddock (London: Trojan Records, 1989).

2. Augustus Pablo, "Tubby's Dub Song," in *King Tubby's in Fine Style*, ed. Osbourne Ruddock (London: Sanctuary Records Group, Ltd., 2004).

3. Karlheinz Stockhausen, "Mikrophonie," in *Mikrophonie I* (Germany: CBS, 1966).

4. The reader is strongly advised to listen to the above recordings since they will be referred to throughout this chapter.

5. Stephen Hill, *The Tragedy of Technology* (London: Pluto, 1988), 3.

6. Karlheinz Stockhausen, *Kontakte* (Germany: Wergo, 1960).

7. The original five-channel version was never performed mainly due to the impracticability of suspending the fifth loudspeaker directly above the center of the audience.

8. See Hugh Davies, *Répertoire International des Musiques Electroacoustiques: International Electronic Music Catalogue, Electronic Music Review* Vol. 2 and 3 (Trumansburg, NY: 1969).

9. Rolf Gehlhaar, interviewed May 2011.

10. Stockhausen, *Mikrophonie I: Für Tamtam, 2 Mikrophone, 2 Filter und Regler, Nr. 15, 1964: 6 Spieler* (London: Universal Edition, 1974).

11. Ibid.

12. Stockhausen, *Kontakte Nr. 12: Elektronische Musik; Kontakte Nr. 12: Für Elektronische Klänge, Klavier und Schlagzeug* (London: Universal Edition, 1968).

13. Lloyd Bradley, *Bass Culture: When Reggae Was King* (London: Viking, 2000), 157.

14. Grover C. "Jeep" Harned, "Interview with Jeep Harned—the Man behind MCI." http://www.mcirecording.com/index1.php?id=2 (accessed September 28, 2012).

15. King Jammy quoted in Michael E. Veal, *Soundscapes and Shattered Songs in Jamaican Reggae* (Middleton, CT: Wesleyan University Press, 2007), 114.

16. Chris Lane, "A Musical Revolution," *Natty Dread*, 2003.

17. Chris Lane, interviewed July 2009.

18. Examination of these filters was made possible by the generosity of Volker Müller.

19. "W49 Hörspielverzerrer" (Germany: Telefunken).

20. "Verzerrer für Hörspielzwecke" (Nordwestdeutscher Rundfunk, 1951).

21. Operated in conjunction with the shop Musikinstrumente & Design accessible at http://www.vintageaudioberlin.de/vabmenue1/vabmenue3.htm.

22. Verzerrer für Hörspielzwecke.

23. W49 Hörspielverzerrer.

24. Stockhausen, *Mikrophonie I: Für Tamtam, 2 Mikrophone, 2 Filter und Regler, Nr. 15, 1964: 6 Spieler.*

25. This technique can be observed in the footage of Prince Jammy mixing a live dub of "Jailhouse Rock" from Howard Johnson and Jim Pines, *Reggae: Deep Roots Music* (London: Proteus in association with Channel Four Television, 1982). This is viewable here http://www.youtube.com/watch?v=kdpuJRpdheA (accessed August 30, 2011).

26. Osbourne Ruddock, *King Tubby's in Fine Style*, ed. Dave Hendley (London: Trojan, 2004).

27. Rolf Gehlhaar, interviewed May 2011.

28. Stockhausen, *Mikrophonie I: Für Tamtam, 2 Mikrophone, 2 Filter und Regler, Nr. 15, 1964: 6 Spieler.*

29. Johnson and Pines, *Reggae: Deep Roots Music.*

30. Lane was accompanied on this trip by Dave Hendley, who compiled the *King Tubby's in Fine Style* CD.

31. Johnson and Pines, *Reggae: Deep Roots Music.*

32. Osbourne Ruddock, *Dub Gone Crazy* (Manchester, Blood and Fire, Ltd., 1994).

33. References to King Tubby's "Big Knob" filter proliferate on internet dub forums as well as in many books and articles about Jamaican music.

34. This is beautifully exemplified in the Korg MS10 and MS20 synthesizers designed in the late 1970s, which had knobs for adjusting the filter frequency twice the diameter of the knobs for nearly all the other parameters.

35. Bradley, *Bass Culture: When Reggae Was King*, 316.

36. Holness, "Rebel Dance."

37. Stockhausen, *Mikrophonie I: Für Tamtam, 2 Mikrophone, 2 Filter und Regler, Nr. 15, 1964: 6 Spieler.*

38. Rolf Gehlhaar, interviewed February 2009.

39. Karlheinz Stockhausen and Jerome Kohl, "Electroacoustic Performance Practice," *Perspectives of New Music* 34(1) (1996).

40. Veal, *Soundscapes and Shattered Songs in Jamaican Reggae*, 114.

41. Stockhausen, *Mikrophonie I: Für Tamtam, 2 Mikrophone, 2 Filter und Regler, Nr. 15, 1964: 6 Spieler.*

42. It is also worth pointing out that the switching mechanism of the W49 does not consist of carbon strips, but that there are carbon strips and wipers in the W66C faders bolted to the side of the W49.

43. There are parts of some of the *Klavierstücke* where the notation is deliberately made so difficult as to provoke the player into playing in a certain way rather than playing exactly what is written.

44. Rolf Gehlhaar, interviewed February 2009.

45. Bradley, *Bass Culture: When Reggae Was King*, 316.

46. Stockhausen and Kohl, "Electroacoustic Performance Practice."

47. Lane, "A Musical Revolution."

48. Stockhausen and Kohl, "Electroacoustic Performance Practice."

49. Ibid.

50. Kevin Ryan and Brian Kehew, *Recording the Beatles: The Studio Equipment and Techniques Used to Create Their Classic Albums* (Houston, TX: Curvebender Publishing, 2006).

51. Ibid., 91.

Bibliography

Bradley, Lloyd. *Bass Culture: When Reggae Was King*. London: Viking, 2000.

Davies, Hugh. *Répertoire International des Musiques Electroacoustiques: International Electronic Music Catalogue, Electronic Music Review* Vol. 2 and 3. Trumansburg, NY: 1969.

Harned, Grover C. "Jeep." "Interview with Jeep Harned—the Man behind MCI." http://www.mcirecording.com/index1.php?id=2 (accessed September 28, 2012).

Hill, Stephen. *The Tragedy of Technology*. London: Pluto, 1988.

Holness, Winston "Niney the Observer." "Rebel Dance." In *King Tubby's Special 1973–1976*, ed. Osbourne Ruddock. London: Trojan Records, 1989.

Johnson, Howard, and Jim Pines. *Reggae: Deep Roots Music*. London: Proteus in association with Channel Four Television, 1982.

Lane, Chris. "A Musical Revolution." *Natty Dread*, 2003.

Pablo, Augustus. "Tubby's Dub Song." In *King Tubby's in Fine Style*, ed. Osbourne Ruddock. London: Sanctuary Records Group Ltd., 2004.

Ruddock, Osbourne. *Dub Gone Crazy*. Manchester: Blood and Fire Ltd., 1994.

———. *King Tubby's in Fine Style*, ed. Dave Hendley. London: Trojan, 2004.

Ryan, Kevin, and Brian Kehew. *Recording the Beatles: The Studio Equipment and Techniques Used to Create Their Classic Albums*. Houston, TX: Curvebender Publishing, 2006.

Stockhausen, Karlheinz. "Kontakte." Germany: Wergo, 1960.

———. *Kontakte Nr. 12: Elektronische Musik; Kontakte Nr. 12: Für Elektronische Klänge, Klavier und Schlagzeug*. London: Universal Edition, 1968.

———. "Mikrophonie." In *Mikrophonie I*. Germany: CBS, 1966.

———. *Mikrophonie I: Für Tamtam, 2 Mikrophone, 2 Filter und Regler, Nr. 15, 1964: 6 Spieler*. London: Universal Edition, 1974.

Stockhausen, Karlheinz, and Jerome Kohl. "Electroacoustic Performance Practice." *Perspectives of New Music* 34(1) (1996): 74–105.

Veal, Michael E. *Soundscapes and Shattered Songs in Jamaican Reggae*. Middletown, CT: Wesleyan University Press, 2007.

"Verzerrer für Hörspielzwecke." Nord West Deutche Rundfunk, 1951.

"W49 Hörspielverzerrer." 4. Germany: Telefunken, undated.

Video

Johnson, Howard, and Jim Pines, "Reggae: Deep Roots Music." London: Proteus in association with Channel Four Television, 1982.

CHAPTER 7

The Oramics Machine and the Origins of British Electronic and Computer Music

Mick Grierson and Tim Boon

The patent office said "the whole concept is too advanced—no one in industry would be interested" . . . looking back at it now, it was quite a ground-breaking concept, and it was all due to Daphne's childhood dream of wanting to do everything with a pen and pencil . . . I'd like to think that if it hadn't been for us, the idea [of graphical timelines] wouldn't exist."

—GRAHAM WRENCH, 2010[1]

IN JULY 2011, the Science Museum placed on display the Oramics Machine, a unique relic of the pioneering years in the 1960s when, arguably, electronic music was being invented. This is the first fruit of a partnership between the Science Museum and Goldsmiths, University of London—a joint research, conservation, and display project to make publicly accessible the work of its inventor, Daphne Oram. After many decades of neglect Oram (1925–2003) is becoming recognized as a highly significant figure in the development of British experimental electronic music. Her pioneering approach to creative engineering, electronic music, and composition made hers a distinctive contribution. By virtue of the technical means she developed, she created sounds and music that we now recognize as quintessentially British—quirky, unsettling, unique, and eerie sound worlds that bring to mind the science fiction and technological heritage of the 1960s. But among today's experimental sound art and electronic music practitioners and scholars, she is gaining greater recognition both as a pioneer of electronic music and sound synthesis techniques and as a composer of the first order, who can also be thought of as the forgotten originator of much electronic music and sound composition technique. For these people, the similarity of the graphical interface of the Oramics Machine programmer to contemporary computer music software such as Cubase is particularly striking. As Peter Manning stated at a symposium on Oramics in 2008, the huge international significance of what she achieved has still not been fully recognized.[2]

Although largely forgotten at the time of her death in 2003, what remained of her reputation was sufficient to justify the preservation of her legacy. Goldsmiths, in close collaboration with the Sonic Arts Network (now soundandmusic.org) and the Oram estate, acquired her collected works and professional effects so that they might be more adequately cared for by the college's Special Collections Department.[3] Staff associated with the Department of Music, including those working in and around the electronic music studio, set about the preservation and cataloguing of the contents of the collection, including recordings, blueprints, computer disks and equipment, papers, letters, and teaching materials. The collection was further developed with the aid of Arts and Humanities Research Council (AHRC) funding, and as a result, the collection is now an accessible research resource. The collection raises significant questions regarding the development of electronic music technology in this period, yet the entire resource, including an enormous amount of mid-twentieth-century music by Daphne Oram, remains as yet entirely unheard, and significantly underresearched.

In 2008, investigations revealed that the Oramics Machine, the unique graphic sound device with which she created music by painting on transparent media, had survived. The electroacoustic musician and writer Hugh Davies had been charged with finding a home for the machine when Oram had been obliged to leave her home and studio at Tower Folly following a stroke. He arranged for it to be given to the Museum of Synthesizer Technology in Hertfordshire, a privately funded museum set up by Martin Newcomb, who exhibited part of the machine for some years. Peter Forrest, author of *The A–Z of Analogue Synthesisers*, acquired the machine from here at auction and, after a period of keeping it in his own studio, he transported it for safekeeping to his property in France.[4] Mick Grierson, representing Goldsmiths Department of Music, with crucial assistance from Carolyn Sturdy and Jo Hutton, acquired it from there in 2008, and in collaboration with the Daphne Oram Trust approached the Science Museum. The museum enthusiastically agreed to add the Oramics Machine to the national collection, on the grounds that its significance, interest, and display potential easily justified the significant conservation attention required to stabilize it. Since then, we have been working together to preserve, display, and understand this find both in terms of its operation and also of its historical significance, while actively seeking funding to enable the project to continue. Together we are supervising a collaborative doctorate to study and contextualize the machine; what follows is therefore a provisional account drawn from preliminary investigations conducted since the college's preservation of the archives, and also in preparations to display it at the museum in a temporary exhibition, *Oramics to Electronica: Revealing Histories of Electronic Music* (2011–2012). In the longer project, we hope to address the seeming paradox that something seemingly so significant had so readily passed out of the public eye; part of that will be to understand more of the bulk of Oram's career from 1959 onward (Figures 7.1, 7.2, and 7.3).

Daphne Oram and the History of Electronic Music

Experimental electronic music and composition is often associated with the postwar commercialization of the tape machine, and Bob Moog's voltage controlled synthesizer.[5] It is well known that these synthesizers were preceded by a range of electromechanical and electronic musical instrument designs, including Thaddeus Cahill's Telharmonium (1897) and Luigi Russolo's Intonarumori (1913).[6] In addition, state-funded electronic music and graphic sound experimentation featured in the Soviet Union throughout its

FIGURE 7.1
The Oramics Machine "programmer" during exhibition installation, with replica film strips made by the artist Aura Satz. Courtesy of the Science Museum (Inv. 2010-68). Images available from Science & Society Picture Library.

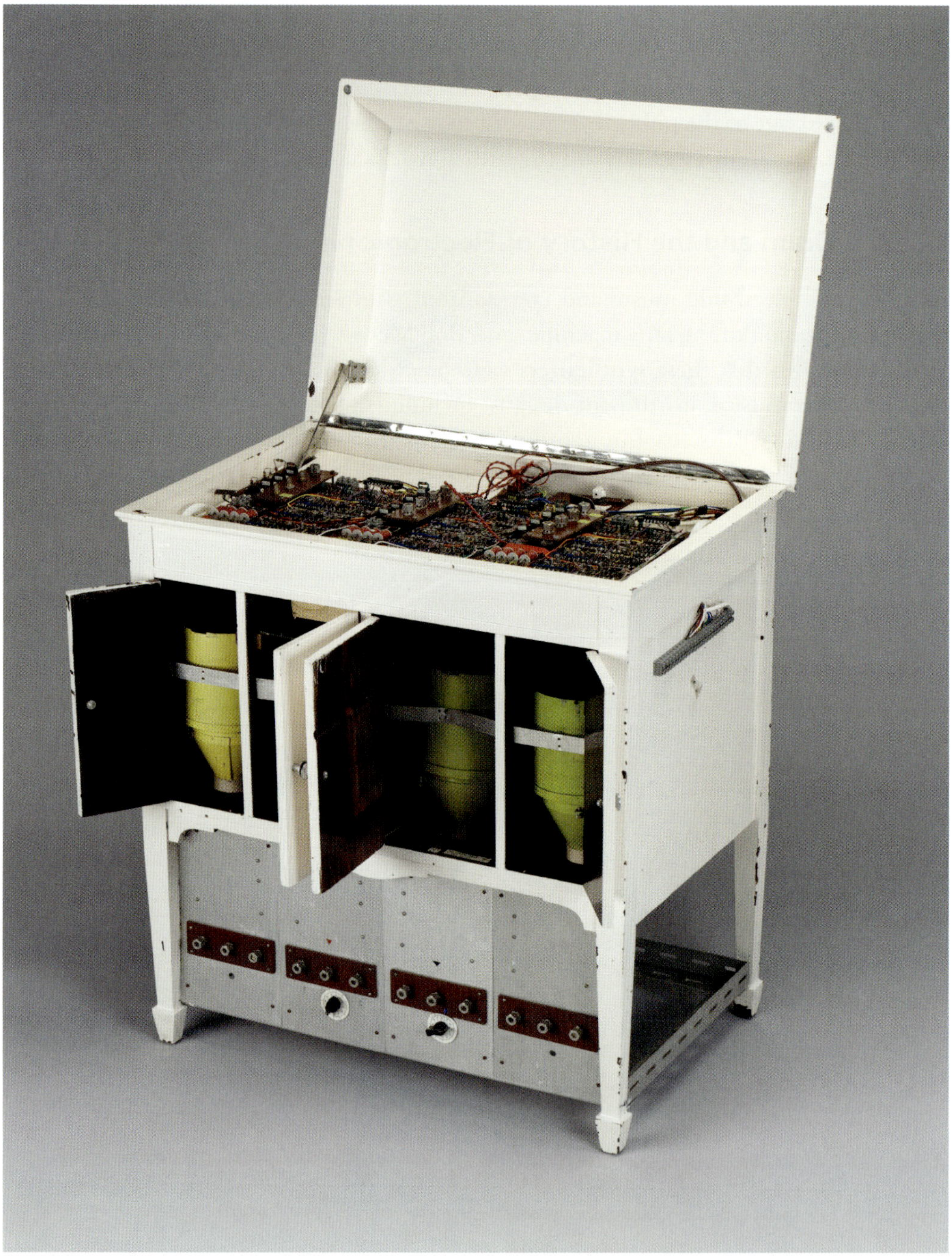

FIGURE 7.2
The Oramics Machine tone generator cabinet. The spaces for four cathode ray tube waveform scanners are clearly visible in the lower part of the picture. Courtesy of the Science Museum (Inv: 2010-68). Images available from Science & Society Picture Library.

FIGURE 7.3
Two waveform slides painted by Daphne Oram. Courtesy of the Science Museum (Inv: 2010-68). Images available from Science & Society Picture Library.

early period. The most famous example is the Theremin,[7] upon which the Ondes Martenot is partly based.[8] Lesser-known Soviet technology focused on graphic representations of sound, of which there are some staggeringly complex examples.[9] This approach relied on early film technology for sonic reproduction. In Austria, Emerich Spielmann developed his "superpiano," which was based on sounds on celluloid discs read like a film's optical soundtrack.[10] Elsewhere, filmmakers and artists including Oskar Fischinger, Rudolf Pfenninger, John and James Whitney, Norman McLaren, and Len Lye experimented with methods of sound synthesis using similar optical processes as a form of abstract animation.[11]

In the various accounts of the development of electronic and computer music,[12] postwar electronic music and sound art is thought of as developing alongside the instigation of two European centers—the Office de Radiodiffusion-Télévision Française (RTF) in Paris, and what became the Westdeutscher Rundfunk (WDR) in Cologne.[13] These two centers are most often associated with the work of the composers Pierre Schaeffer and Karlheinz Stockhausen, respectively. In the United Kingdom, the BBC Radiophonic Workshop represents a similar attempt to respond to the combination of technological and cultural factors creating a revolution in sound and music. Daphne Oram had a key role in its foundation.[14]

Early in her career Oram had declined a place at the Royal College of Music to become a music balancer at the BBC. Throughout the Second World War she worked as an audio engineer, standing by during live concert performances to "drop in" recorded versions of music in the event of an air raid. Following this period, in the years leading up to 1958, Oram, who was acquainted with the French and German experiments, took to working overnight, assembling and disassembling equipment on a daily basis at the BBC, to create a makeshift studio—a place where she could develop her experimental electronic music technique. This led to a number of small projects with the BBC Drama department, including *Private Dreams and Public Nightmares* (1958), demonstrating the power and potential of "Radiophonics"—a term adopted in Britain to represent electronic music and sound practice, especially where used as an adjunct to radio and, later, television. It appears that Oram's experiments—along with the efforts of Desmond Briscoe—were the genesis of the BBC Radiophonic Workshop. As cofounder of the Workshop, Oram invented and refined a number of key working practices that became central to sound design and electronic music production (the use of multiple tape loops for pitched electronic music, for example). This, it can be argued, directly influenced the work of significant electronic musicians and engineers in the United Kingdom during this period, including Delia Derbyshire and others at the Workshop.[15]

Oram left the Radiophonic Workshop and the BBC in late 1958, disappointed that it did not intend to pursue the musical line she had hoped for. She immediately set about establishing a private electronic music studio at Tower Folly, the converted oasthouse in Kent where she lived.[16] Here she produced some music for commercial clients (including Lego and Rotolock valves), ran courses where she taught musique concrête technique, and set about developing her ideas about audiovisual composition, which resulted in the development of the Oramics Machine. Perhaps she saw her studio as a latter-day implementation of aspects of Solomon's House from Francis Bacon's seventeenth-century utopia, *The New Atlantis*, a pencil-annotated copy of which survives in the collection at Goldsmiths:

> We have also sound-houses, where we practise and demonstrate all sounds, and their generation. We have harmonies which you have not, of quarter-sounds, and lesser slides of sounds. Divers instruments of music likewise to you unknown, some sweeter than any you have, together with bells and rings that are dainty and sweet. We represent small sounds as great and deep; likewise great sounds extenuate and sharp; we make divers tremblings and warblings of sounds, which in their original are entire.

In the early days of the studio, she wrote an updated version of this vision, which she titled "Atlantis Anew," a piece that alluded to both music therapy and to esoteric aspects of music.[17]

The Oramics Machine

In addition to the establishment of one of the first electronic music studios, Oram is credited with the invention of a prototypical form of sound synthesis in the machine, a system she referred to as *Oramics* (although, confusingly, she used the term to describe any music that she created). The machine represents a fascinating alternative path in the development of electronic music when compared with the more familiar routes via voltage control and computer techniques that were under development at the same time. Not only was this one of the earliest forms of electronic sound synthesis and control, it differed from most other contemporary electronic music devices by virtue of its audiovisual input technique. In this, the composer draws onto a synchronized set of ten 35 mm film strips that run across the top of the "programmer"; these are read by a series of photoelectric transistors, generating electrical charges to control frequency, amplitude, timbre, and duration. As a variant use of film technology, it has something in common with both European and American film art,[18] and the optical sound experiments in Soviet Russia of the 1930s mentioned above.[19] Our first investigations have recently shown that Oram took these principles much further than any of her contemporaries.

She received two grants from the Gulbenkian Foundation to develop the Oramics Machine, first in 1962, and then again in 1965. This enabled her to enlist the help of a number of engineers, including her brother (who designed and built the servo mechanisms) and two electronics engineers. During the first award, the telephone engineer Fred Wood assisted in developing the Oramics Machine; the device was fully realized by the talented engineer Graham Welch, who had worked on radar during his four years' of service with the Royal Air Force. He created a reliable and flexible sound generation and control mechanism, possibly based on an earlier prototype.[20]

Through discussions with Wrench, we have come to a basic understanding of the machine's capability. If substantiated by further research, this may support Daphne Oram's claim that the system was, in a sense, the first British music computer, and the first graphic sound composition device, in that it is a digitally controlled sound synthesizer that can be operated by organizing graphical information on a time line. The machine was developed at a time when there were no widely used commercial sound synthesizers, and long before electronic devices had become a mainstay of communications and personal media. Moog's paper on voltage-controlled synthesizers was not published until 1964, and the prototypes were only unveiled later that year.[21] However, the Oramics Machine was not just an attempt to build a synthesizer. Rather, it was an attempt to create a vision for new music—a vision to make possible and drive forward the further exploration of electronic music without necessarily adhering to existing traditions, frameworks, languages, and structures of music. For Daphne Oram, it seems, all of these conventions, while being of importance, did not necessarily preclude the creative and technical study of new developments in music and sound. The vision and forethought behind the Oramics Machine goes well beyond what one would expect from an instrument designed and built to carry out esoteric composition. According to Wrench, Oram wanted to create a system on which composers would be able to both imagine and draw the "whole of the music simultaneously." Her vision was to translate visual and graphic processes—shapes, curves, images, and their encapsulated gestures—into sonic compositional elements. This distinguishes her approach from that of contemporaries, including Moog, Moog's associate Raymond Scott,[22] and Peter Zinovieff,[23] even the creators of the RCA synthesizer, all of whom in different ways and with different emphases were exploring both sonic and compositional aspects of the new field of electronic music.

Essentials of the Oramics Machine

Two connected aspects of the machine reveal its significance; that its pitch was digitally controlled and that its sound generation and control processes—including both waveform generation and sound qualities—were graphically/optically controlled. The machine was not tuned to equal temperament, as is the case with most musical instruments. Rather, the frequency range was expressed and set in increments of single frequencies. The waveform scanning device designed by Wrench was capable of ranging from well below 20 Hz to just below 10,000 Hz, stepped at one cycle per second. This was achieved through the implementation of four separate frequency control tracks, each with its own control mechanism. These were the first four film strips on the Oramics interface, starting from the top, going down. Each strip was used to set a value between 0 and 9 (although as it was a four-bit system, it would have theoretically been capable of setting values ranging from 0 to 15). The first strip sets individual units of frequency from 0 to 9; the second strip from 10 to 90; the third from 100 to 900; and the last from 1,000 to 9,000. This allowed for the frequency to be set by hand as described, at single frequency intervals up to 9,999 Hz through a four-bit binary representation (Figure 7.4).

This value was used to set the timebase for the waveform scanning system, which in turn "read" a waveform drawn onto a glass slide. The timebase frequency controls were precisely calibrated components switched in by coil-actuated reed switches. Graph paper on the top of the "programmer" was critical to make sure that the frequency tracks were properly aligned. The waveform scanning system was of particularly innovative design. In the early iterations of the machine, the scanners were housed in an old government surplus steel cathode ray tube (CRT) framework. An electron multiplier tube with an amplifier and feedback circuit was set up in front of an oscilloscope's CRT, surrounded with a dark cloth and dark paper. The success of the technique was dependent on the scanner being able to reset itself to the beginning of the waveform as soon as it had reached the end of the cycle. This was a significant problem; most oscilloscopes were unable to get back to the next scanning cycle fast enough to create a continuous audible waveform (Figures 7.5 and 7.6). According to Wrench, the Oramics system deployed a technique common to RAF radar technology:

FIGURE 7.4
Oram working at the Oramics Machine (ca. 1969); the top three films with masks of paper or tape controlled the pitch; the other films controlled the quality of the sound. Copyright © Daphne Oram Archive, Goldsmiths, University of London.

FIGURE 7.5

Circuitry of the frequency control scanners in the top of the waveform unit. Several sets of coil-actuated reed switches (for pitch) can be seen. Courtesy of the Science Museum (Inv: 2010-68). Images available from Science & Society Picture Library.

FIGURE 7.6
Detail of filament lamps; the photo transistors are mounted in the dark strip below the film strips. Courtesy of the Science Museum (Inv: 2010-68). Images available from Science & Society Picture Library.

> The timebase is . . . (not) allowed to free range back which is always an exponential wave form. We actually drove the fly back in the same way that we drove the initial scan—made it very linear. We made it very linear but fast . . . at least 10 times higher than the highest audible frequency we wanted. The flyback was audibly instantaneous. As you repeated the scanning, you never got the audible click. We could scan a sinewave and it was very very pure.[24]

Further research is required to establish the precise history of the machine's construction and subsequent modification as Oram sought to realize her vision. Wrench, over the eighteen months of the second Gulbenkian grant, created a single scanner, parametric controls, and the ability to write a single melody or rhythm line. This principle was designed to be extended to create a fully operational composition system, an ambition that was realized in later years via various modifications that were probably undertaken by another engineer after Wrench's period working with Oram.[25]

To realize Oram's ambition for graphic control of the machine's timbre—amplitude and reverberation levels and vibrato—Wrench developed a sophisticated electronic feedback system. Originally, Wrench on Oram's behalf searched for a method to create a theoretical dynamic range between –60 and 0 dB for the amplitude control, a range unheard of in most audio systems of the time. To our knowledge, there were no wide-range electronic voltage controlled amplifiers (VCAs) available during this period, even in commercial synthesizer products such as the Moog (these types of controls came later). Some motorized faders existed, but they were very noisy and so unusable as part of a composition system. The challenge was to create a continuous varying signal from a shape drawn on a strip of 35 mm film, one that was capable of controlling the amplitude of the signals produced by the waveform scanner. Wrench applied his knowledge of electronics, using an electronic bridging circuit to overcome problems innate to the audiovisual input method.[26] This hugely innovative and resourceful approach gave Wrench and Oram the precision control they needed. It allowed them to implement fully automatable envelopes to control the output of the waveform generators, the amplitude of reverberation signals, and even the precise variations in pitch that gave the Oramics Machine its distinctive and eerie vibrato.[27]

Displaying the Oramics Machine

Within a year of the museum's acquisition of the machine, an opportunity arose to place it on display in the exhibition *Oramics to Electronica: Revealing Histories of Electronic Music*. The way in which this project is framed has enabled us to take into account several views of Oram and her machine, placing them in a longer historical context. Within the context of ambitious plans to develop *Making Modern Communications*, a major gallery due to open in 2014, the museum established its "public history project." The aim of this initiative is to explore how various potential audience groups understand the history of technology. The Oramics Machine was made the heart of a cocreation experiment on the history of electronic music, in which four different groups have been able to develop their own accounts of this history. Individuals have been invited to attend a series of workshops where, with the help of various stimuli—including showings of archive television programs and visits to the museum's reserve collections—they develop narratives based on their own understanding and responses.[28] The process was filmed by the documentarists Nick Street and Jen Fearnley.[29] We collaborated with a selection of people who worked at the BBC Radiophonic Workshop and at Electronic Music Studios in the 1960s and 1970s; their contribution is an autobiographical account of their respective organizations in a showcase each. These are two British examples of developments in electronic music that paralleled Oram's endeavors. The museum has also worked with a group of twelve people involved in electronic music today, who were recruited via the project's Oramics Machine Facebook page; they have produced three thematic showcases on the history of their field. The first of these covers "sonic frontiers," which considers the search for new sounds and novel compositional techniques in electronic music. A second case, "make do and mend," shows examples of the role of DIY, in which musicians have made their own instruments; improvised by using nonmusical devices; or used devices in a way not intended, as in the case of the TB303 bass sequencer designed to accompany soloists and used as the foundation of the Acid House genre. "Democratization" is their final theme, showing how this kind of music has increasingly become accessible to those of slender means. All themes came out of discussions prompted by the case of Daphne Oram, and at least the first two of them explicitly picked up on aspects of the Oramics Machine. We have also worked with a group of women amateur writers, who produced monologues stimulated by the story of Daphne Oram, and with the National Youth Theatre Acting Up 2 access course, who produced *Oramix*, a performance about Oram and their response to her music in the museum in April 2011. Extending the circle of curation, we also commissioned the artist Aura Satz to make a film about the machine (Figure 7.7), *Atlantis Anew.*[30]

Conclusion

The reemergence of the Oramics Machine alters the commonly accepted story of electronica. It is undoubtedly a key relic of the 1960s era of electronic music when several different individuals, groups, and companies were all seeking ways and means to create music that could go beyond the constraints of composition based on tape manipulation. The development of the exhibition in the company of several groups of collaborators has also opened up new vistas and contexts, both for what was already known about Oram and her work, and for the broader field, too. Electronic music is now old enough to have a rich, and contested, history. Within the research project that is only just beginning, we intend to place Oram and her invention properly within that broader context. Along the way we hope to address the nagging question for today's computer musicians of

FIGURE 7.7
The Oramics Machine on display in the *Oramics to Electronica* exhibition, October 2011. Courtesy of the Science Museum (Inv: 2010-68). Images available from Science & Society Picture Library.

why there is such a strong resemblance between the graphical notation of contemporary computer music systems, such as the Cubase Key editor found in Apple's GarageBand software, and the graphical interface of the Oramics Machine programmer.

Notes

1. Graham Wrench (recording), in conversation with Jo Hutton and Chris Weaver, 2010.

2. P. D. Manning and N. Candlish, "Daphne Oram: A Perspective on Her Pioneering Contributions to Technology and Music, with Special Reference to the Unique Circumstances That Shaped and Influenced Her Achievements during Her Time at the BBC," Oramics Symposium, South Bank Centre, London, 2008. Louis Niebur, *Special Sound: The Creation and Legacy of the BBC Radiophonic Workshop* (Oxford, UK: Oxford University Press, 2010), 72–74.

3. From the 1970s onward, Oram entered a period of decline, although the precise details of this period have yet to be investigated in detail.

4. Peter Forrest, e-mail to Tim Boon, October 10, 2011. P. Forrest, *The A–Z of Analogue Synthesisers* (Exeter, UK: Susurreal Publishing, 1998).

5. P. D. Manning, *Electronic and Computer Music: Revised and Expanded Edition* (New York: Oxford University Press, 2004), 102-20.

6. S. Emmerson, *Living Electronic Music* (Farnham, UK: Ashgate, 2007), 135, 149.

7. A. Smirnov and L. Pchelkina, "*1917–1939. Son Z/Sound in Z,*" *PALAIS/Palais de Tokyo Magazine* 7, Paris, 2008, 66–77. See also Albert Glinsky, *Theremin: Ether Music and Espionage* (Urbana: University of Illinois Press, 2005).

8. K. D. Skeldon, L. M. Reid, V. McInally, B. Dougan, and C. Fulton, "Physics of the Theremin," *American Journal of Physics* 66(11) (1998): 945–55.

9. Smirnov and Pchelkina, "*1917–1939.*"

10. Donhauser, this volume.

11. A. Rees, *A History of Experimental Film and Video* (London: BFI, 1999) 28, 35, 58. B. Evans, Foundations of a Visual Music," *Computer Music Journal* 29(4) (2005): 11–24.

12. T. Holmes, *Electronic and Experimental Music: Pioneers in Technology and Composition* (London: Routledge, 2002); J. Chadabe, *Electric Sound: The Past and the Promise of Electronic Music* (Upper Saddle River, NJ: Prentice Hall, 1997).

13. Manning, *Electronic and Computer Music*, 19–73.

14. Niebur, *Special Sound*, 53–73.

15. Niebur, *Special Sound*, 59–61.

16. Hugh Davies, "Daphne Oram (obituary)," *The Guardian*, Friday, January 24, 2003. http://www.guardian.co.uk/news/2003/jan/24/guardianobituaries.artsobituaries (accessed October 16, 2011).

17. This is quoted in Dan Wilson, "The Woman from New Atlantis," *The Wire* 330, August 2011, 29–35.

18. A. Rees, *A History*, 15–75.

19. Smirnov and Pchelkina, "*1917–1939.*"

20. Wrench, recording, Wrench e-mail to Tim Boon, October 19, 2011. The extent, nature and capacities of any early prototypes have yet to be established; recollections of participants differ. Alan Sutcliffe, later of Electronic Music Studios, recalls using such a device in a composition named *Icarus* on a course at Tower Folly in 1958 or 1959, conversation with Tim Boon, September 2011. A recording of this composition is now in the Science Museum EMS collection.

21. T. J. Pinch and Frank Trocco, *Analog Days: The Invention and Impact of the Moog Synthesizer* (Cambridge, MA: Harvard University Press, 2002), 276–301.

22. Jeff E. Winner and Irwin D. Chusid, "'Circle Machines and Sequencers': The Untold History of Raymond Scott's Pioneering Instruments," 2009. http://raymondscott.com/em.html (accessed July 25, 2010).

23. Manning, *Electronic and Computer Music*, 209–13.

24. Wrench, recording.

25. From looking at current photographs of the machine taken since it was adopted by the Science Museum, Wrench has confirmed that a great deal of work was done on it following his period of employment. His view is that "to my knowledge Wood was very able to build circuits as published in various amateur magazines, and indeed built several small pre-amplifiers for Daphne from published articles or kits, but I am not aware of any original designing"; Wrench, e-mail to Boon, October 19, 2011.

26. Wrench implemented a system whereby, through the use of a photocell or photodiode, the sensitivity of one of the arms of a Wheatstone bridge could be varied. When the bridge was balanced, it would output a steady signal at zero. It could then be unbalanced by the photocell, and this would in turn control the voltage, and hence, the amplitude of the signal. The problem was simply to set the base amplitude and sensitivity. However, this proved very difficult. To operate correctly, the photocell needed to be illuminated by a bulb or lamp. Due to the high sensitivity

ratios between the photocell and the bridge (as high as 1000:1), any slight deviation or deterioration in the bulb's brightness would unbalance the system, causing massive alterations in the signal output. To solve this problem Graham attached the bulb to a separate bridge in order to normalize the output of the bulb at all times. This was read by its own separate photocell. The balanced output of the second bridge was used to adjust for any variation or deterioration in the output of the light and effectively engineer out the errors inherent in the system. This in turn fed back to the main bridge, keeping the amplitude of the signal at a steady rate, eliminating the instability brought about by the enormous sensitivity of the photocell. In this way, not only did the signal become far more controllable and precise, most important, when the system was supposed to be at zero, it remained at zero.

27. The vibrato functioned by adding and subtracting from the amplification on the timebase unit itself. This meant that the timebase would be modulated, slowing down and speeding up the scanning rate of the waveform generators. This was simply an extension of the problem that Wrench had solved with the double electronic bridging system, and it shows off the modularity of the system's design.

28. An initial account of the philosophy of this project can be found in T. Boon, "Introduction: Co-Curation and the Public History of Science & Technology" (introduction to a selection of papers from Science Museum International Workshop), *Curator* 54(4) (October 2011): 383–87.

29. A synoptic film is available to view online via the museum's exhibition web pages.

30. This film is also available online to view online via the museum's exhibition web pages.

Bibliography

Bacon, Francis. *Sylva Sylvarum, or, A Natural History in Ten Centuries: Whereunto is Newly Added the History Naturall and Experimentall of Life and Death, or of the Prolongation of Life*, London, 1664.

Boon, T. "Introduction: Co-Curation and the Public History of Science & Technology" (introduction to a selection of papers from Science Museum International Workshop). *Curator* 54(4) (October 2011): 383–87.

Chadabe, J. *Electric Sound: The Past and the Promise of Electronic Music*. Upper Saddle River, NJ: Prentice Hall, 1997.

Davies, H. "Daphne Oram" (obituary), *The Guardian*, Friday, January 24, 2003. http://www.guardian.co.uk/news/2003/jan/24/guardianobituaries.artsobituaries (accessed October 16, 2011).

Emmerson, S. *Living Electronic Music*. Farnham, UK: Ashgate, 2007.

Evans, B. "Foundations of a Visual Music." *Computer Music Journal* 29(4) (2005):11–24.

Forrest, Peter. *The A–Z of Analogue Synthesisers*. Exeter, UK: Susurreal, 1998.

Glinsky, A. *Theremin: Ether Music and Espionage*. Chicago: University of Illinois Press, 2005.

Holmes, T. *Electronic and Experimental Music: Pioneers in Technology and Composition*. London: Routledge, 2002.

Manning, P. D. *Electronic and Computer Music: Revised and Expanded Edition*. New York: Oxford University Press, 2004.

Manning, P. D., and N. Candlish. "Daphne Oram: A Perspective on Her Pioneering Contributions to Technology and Music, with Special Reference to the Unique Circumstances That Shaped and Influenced Her Achievements during Her Time at the BBC." Oramics Symposium, South Bank Centre, London, 2008.

Niebur, Louis. *Special Sound: The Creation and Legacy of the BBC Radiophonic Workshop*. Oxford, UK: Oxford University Press, 2010.

Pinch, Trevor, and Frank Trocco. *Analog Days: The Invention and Impact of the Moog Synthesizer*. Cambridge, MA: Harvard University Press, 2002.

Rees, A. *A History of Experimental Film and Video*. London: BFI, 1999.

Skeldon, K. D., L. M. Reid, V. McInally, B. Dougan, and C. Fulton. "Physics of the Theremin." *American Journal of Physics* 66(11) (1998): 945–55.

Smirnov, A., and L. Pchelkina. "*1917–1939. Son Z/Sound in Z.*" Paris: *Palais de Tokyo Magazine* 7, 2008, 66–77.

Wilson, Dan. "The Woman from New Atlantis," *The Wire* 330, August 2011, 29–35.

Winner, Jeff E., and Irwin D. Chusid. "'Circle Machines and Sequencers': The Untold History of Raymond Scott's Pioneering Instruments," (2009). http://raymondscott.com/em.html (accessed July 25, 2010).

Wrench, Graham (recording), in conversation with Jo Hutton and Chris Weaver, 2010.

CHAPTER 8

Musical Instruments and User Interfaces in Two Centuries

Tellef Kvifte

Introduction

In what ways are electronic instruments different from acoustic instruments? Do they constitute a radical break with the past, and, if so, what are the new characteristics more specifically? There are many opinions, along a continuum from regarding the electronic instruments as representing the dawn of a new golden age, where music finally is liberated from the imperfections of acoustic instruments and performers of varying degrees of accomplishment, to absolute rejection, condemning the electronic instruments and their sound as the end of real music as we know it—as the final takeover of the machines. Criticism and praise is directed both on how the instruments are played as well as the sounds they produce.

While it *is* evident that electronic instruments in many important respects differ from acoustic instruments, I am also interested in the continuity between those two categories of artifacts. They are, after all, both characterized as *musical instruments*, and, as such, they obviously have some properties in common. One such property is the ability to produce sound that is useful in a musical context; another that the sound-producing process is possible to *control*. My discussion takes precisely this control aspect as its point of departure, and in the first part of this chapter, I will look principally at the *control organs* of a number of instruments, and not very much at other (equally interesting) aspects of the instruments, like the sound-producing mechanism or materials used in the construction. The control organs are those parts of the instrument that are sensitive to the movements of the performer and form what we also call the *user interface* of the instrument.

To illuminate similarities and differences in user interfaces of acoustic and electronic instruments, I will use two periods in the history of (Western) musical instruments as examples: first, the nineteenth century, with the development of a large variety of free-reed instruments, and of the modern orchestral woodwinds; then, the second half of the twentieth century, with the development of synthesizers, electronic, and computer-based instruments. In both these periods, a large number of new user interfaces were developed, and there was an almost explosive increase in the number of control organs on some instruments. In both instances, sound-producing mechanisms that were previously little used in Western music were introduced, namely free reeds and electronic circuits respectively. But the emphasis in the development of new control organs was different. The nineteenth-century development was mainly concerned with the control of the discrete pitch entities of scale steps, while the late twentieth century development showed an unprecedented interest in the control of timbral qualities, as will be discussed below.

The development of musical instrument user interfaces is part of larger systems of technologies, cultural values, and practices. Toward the end of the chapter, I will try to position the two periods of instrument development in the context of dominating technologies for the distribution of music, namely standard music notation in the first period, and sound recording technology in the second. To understand the nature and development of these technologies, in turn, depends on still larger contexts that we will not be able to say much about in the scope of this chapter.

A study like this depends on the availability of a large number of musical instruments from different periods of history, including instruments that are no longer in use, whether they are regarded as obsolete or never were generally accepted. Ringve Museum in Trondheim, Norway, has a collection that inspires the kind of questions posed in this chapter, and the many types of instruments there afford insights otherwise hard to acquire. Most instruments referred to in this chapter can be found in this excellent collection.

The value of the collection lies not only in the availability of exemplars otherwise hard or impossible to obtain, but, even more, in the perceptions gained from the juxtaposition of instruments from many different contexts. Seeing otherwise unrelated artifacts together in the context of a museum collection encourages new perspectives, and it adds to the understanding of their similarities across space, time, and type, as well as providing a clearer grasp of their individual characteristics.

Nineteenth-Century Interfaces

Free-Reed Instruments

From the beginning of the nineteenth century, the free reed as a sound-producing device was used in a large number of more or less similar instruments that were eventually known as aeolina, accordion, concertina, mouth harmonica, bandoneon, house organ, and harmonium, to name a few. As used in these instruments, the free reed was a very robust and stable sound producer that would keep in tune for a long time with a minimum of maintenance. Whether for this reason, or for its sound quality or for its novelty, instruments based on the free reed gained enormous popularity during the century.

Along came a range of interfaces, as part of various mouth-blown and bellows-driven instruments. Some of the constructions, like many different accordion types, are still in use, but many were short-lived and made only as prototypes or produced in small numbers. Many of these can be found in museums, and, even if they did not come into general use, may have stories to tell and insights to offer. Those specific constructions that did not gain acceptance may sometimes highlight particular themes and problems better than more lasting solutions.

The free reed is, like other types of reed, driven by an air stream that is provided either by direct blowing into the instrument or via some mechanical system of bellows. In the traditional Western single- and double-reed instrument family of clarinets, oboes, and bagpipes, one reed is feeding an air column that provides all pitches available on the instrument through manipulation of the length of a resonating air column. In the typical free-reed instrument developed in the nineteenth century, however, there is one reed for each available (fixed) pitch. In all these instruments, there are in principle two aspects of the user interface: one set of buttons or keys used to select pitches, and a means for providing air that can set the relevant reeds in motion. The latter is either in the form of direct blowing into the instrument (as in the mouth harmonica and the later melodika), hand-operated bellows (as in accordions of different kinds), or by bellows operated by pedals (as in harmoniums or house organs). In most cases, this part of the user interface can also control the loudness of the sound. Finally, some, but not all, kinds of free-reed instruments have some ability to change sound quality by providing a choice between different sets of reeds, or by engaging some mechanical device around the reeds to regulate the passage of the sound.

The pitch control interfaces on free-reed instruments take many forms. Some interfaces are modeled after existing instruments, like the standard piano keyboard. The keyboard was used later in the century on house organs or harmoniums, and also on some varieties of accordions.

The *German Æolian Tutor*, published in 1830 by I. Willis and Co., shows a picture of the "keyed Æolian," and the description of the instrument is mainly centered around pitch control, the available range and, in some detail, the organization of pitches on the layout of the keys:

> The notes of the Diatonic Scale being placed alternately on either side, and the intermediate semitones in the middle, this arrangement enables the performer not only to play simple harmonies, as well as those of the most difficult combinations, and to modulate into various keys, but also to perform an air with a distinct accompaniment, the same as on the piano-forte or other keyed instrument.[1]

Many different button layouts are also to be found, illustrating a blossoming creativity in both technical and musical matters. A number of those are based on a principle in which each button controls two different pitches, depending on the direction of the air—blow or suck on a mouth harmonica, push or pull on a diatonic accordion.[2]

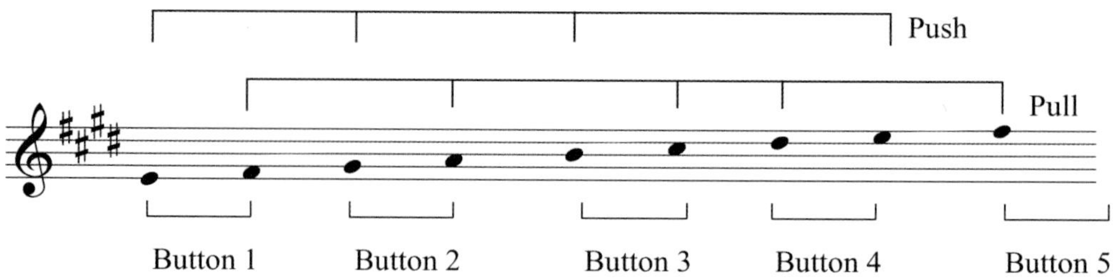

FIGURE 8.1
Layout of pitches on a typical diatonic accordion.

Figure 8.1 shows a typical ordering, where the "push" action on the bellows lets the right-hand buttons produce pitches all within the major triad, while the "pull" action lets the buttons produce the remaining four pitches of the major scale. The left-hand buttons, producing roots and chords, work in a similar fashion, usually producing a tonic chord on push and a dominant chord on pull.

A diatonic accordion may have more than one row of right-hand buttons; two and three are quite common, the additional rows being tuned in a different key. Figure 8.2 shows two rows, tuned a fifth apart, allowing for a playing style with slightly greater flexibility in the relationship between pitch and direction of bellows. In the one-row variety, the root note of the row, for example, will always have to be performed with a push action, while, as can be seen in Figure 8.2, the root note of the inner row can also be performed on the *outer* row with a pull action.

The diatonic button system affords a compact user interface, as each button is used for two pitches, and a basic ordering of the available pitches in the two categories of "root triad" and "the rest." In certain genres this is a sensible and practical setup, with an almost foolproof system for harmonization of melodies.

Also common are varieties of chromatic button accordions. A five-row type is shown in Figure 8.3 (top) and in Figure 8.4. Characteristic of these chromatic button user interfaces is their symmetry; in this example the buttons are ordered in minor thirds vertically and minor seconds diagonally. Playing each diagonal row in turn produces a chromatic scale; to facilitate orientation the buttons of a C major scale are white (Figure 8.5).

Looking at Figure 8.3, it is striking that the buttons are all used for the control of one single parameter in music, namely

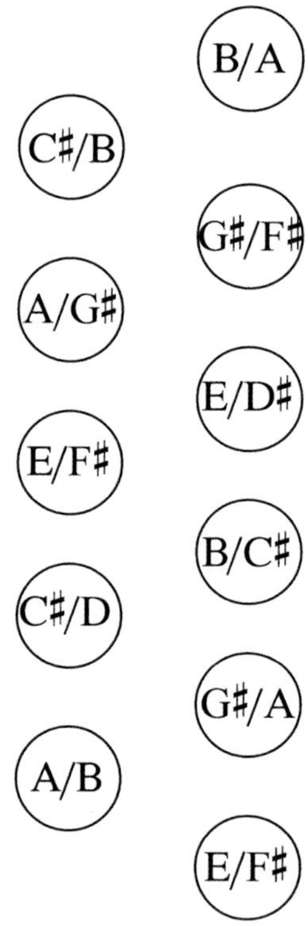

FIGURE 8.2
Relation of buttons to pitches on a two-row diatonic accordion, left letter push, and right letter pull.

FIGURE 8.3
Two varieties of accordion: (top) chromatic five-row and (bottom) two-row diatonic. Photo by Tellef Kvifte. Courtesy of Ringve Museum.

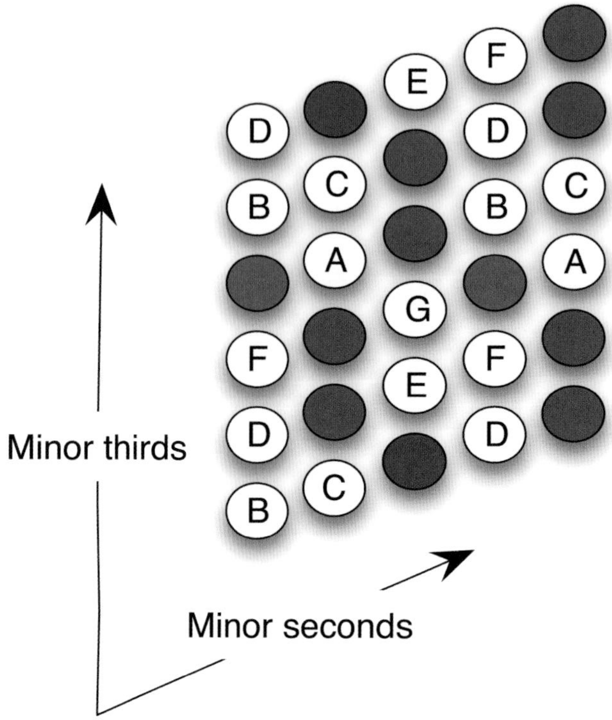

FIGURE 8.4
Mapping of buttons to pitch on a five-row chromatic accordion.

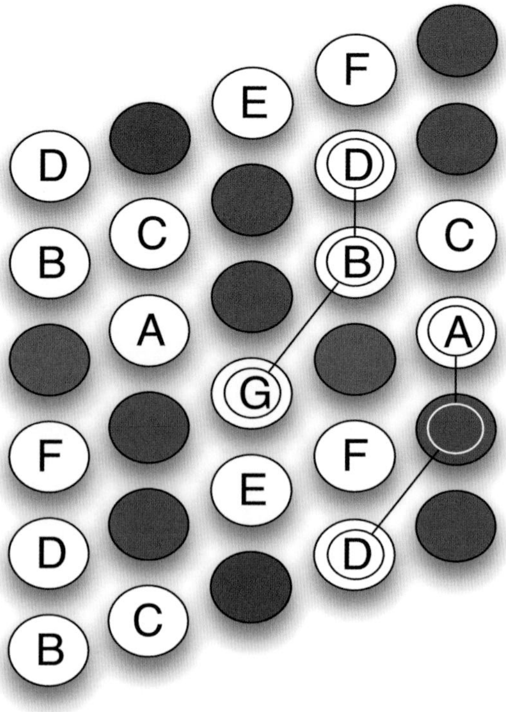

FIGURE 8.5
Patterns for major triads on a chromatic accordion.

pitch in the form of scale steps. Also evident is that, despite the very different layouts, they all relate quite directly to the dominating basic pitch organization characteristic of the Western world, and they are easily connected to the pitch representational system of standard notation. The pattern of black-and-white buttons on the chromatic accordion is also a direct analogy to the standard piano keyboard, or, for that matter, to notes without sharps or flats in standard notation.

Also on the concertina (Figure 8.6), where the button system at first sight seems quite chaotic (Figure 8.7), the analogy is quite close, and in the patent documents from the inventor, Charles Wheatstone describes the system like this:

> The notes of the scale are placed alternately on each side of the instrument; all the notes written on spaces being on the right side . . . and all those written on lines on the left-hand side . . . By this arrangement, to perform a diatonic scale in any key the first and second fingers of both hands only are needed, and no crossing of the fingers ever occurs.[3]

A more fanciful example is the cecilium, loosely based upon the cello (Figure 8.8). Here too, free reeds account for the sound-producing mechanism. Pitch control is

FIGURE 8.6
Concertina by Charles Wheatstone. Photo by Tellef Kvifte. Courtesy of Ringve Museum.

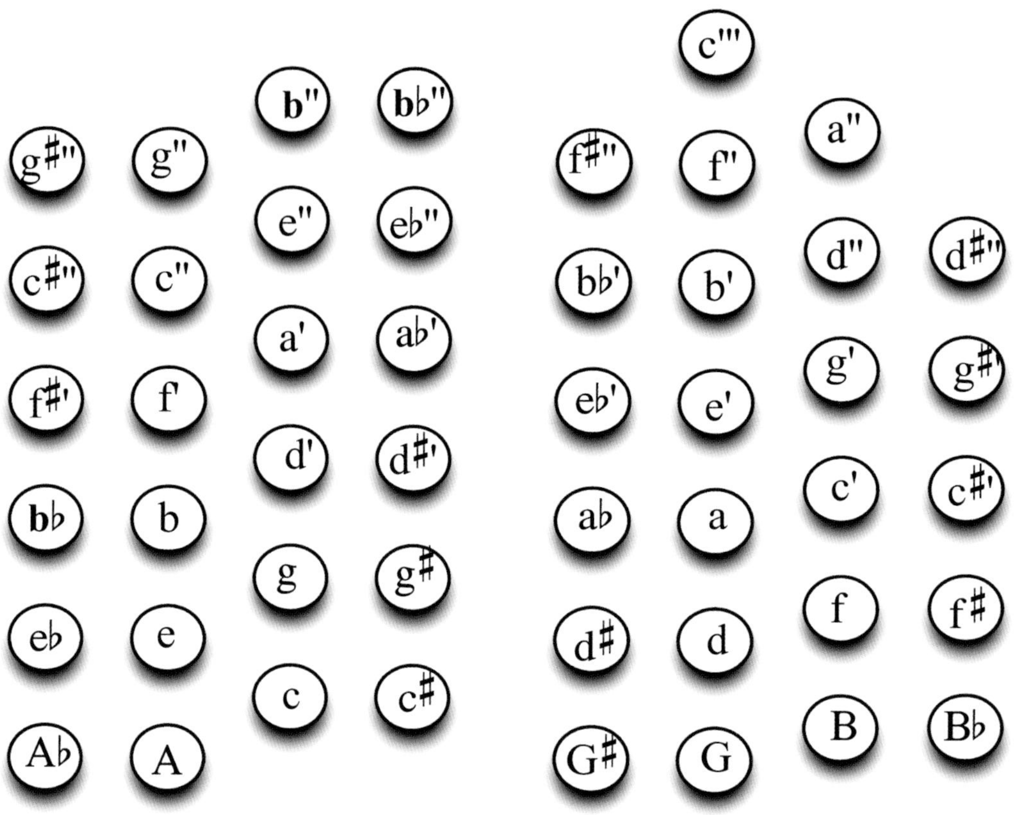

FIGURE 8.7
Button layout of the English concertina.

provided by a set of small buttons, laid out in close analogy to the string fingerboard (Figure 8.9), and there is a handle to be moved back and forth in a position corresponding to the bow of a cello to provide energy for the sound production (by operating the bellows) and thereby also to control loudness.

Even if this instrument never came into general use and is today found almost exclusively in museums, it is an illustration of a certain point in the development of user interfaces. Unlike the aeolian and the accordions described above, the cecilium is modeled after an existing instrument. What is interesting about this is not only that there are some similarities between the cecilium and the cello, but also because there are important differences. On the cecilium, the left hand can produce well-defined pitches, in the same way as is possible on a cello by shortening strings with fingers. Put down the index finger to get one pitch; put down the ring finger to get a different one.

At the same time it is, on a cello, possible to produce any pitch *between* those two—the fingerboard offers the means to produce a continuously variable pitch series. If this had been an indispensable feature of a musical instrument, the cecilium would not have been made. Thus, the cecilium points to the fact that interface development in the nineteenth century is more concerned with the control of discrete pitches (scale steps) than with continuously variable pitch, as in vibrato and glissandi. While the same point can be seen also in the other instruments discussed in this section, it becomes even more obvious in the case of the cecilium, because of the analogy with the cello.

FIGURE 8.8
Cecilium. Photo by Tellef Kvifte. Courtesy of Ringve Museum.

The Woodwinds

Parallel to the invention of the great number of free-reed instruments, there was a radical development in the user interfaces of woodwind instruments, among them the orchestral woodwinds of flute, clarinet, oboe, and bassoon; but they are not confined to these, as can be seen in the construction of new instruments like the saxophone, following similar principles as the orchestral instruments.

The traverso, or transverse flute as used in the Baroque era, may serve as a simplified example of the basic interface of this class of instruments (Figure 8.10 bottom).

The sound is produced by blowing across the hole to the left in Figure 8.10, where the wind energy will excite standing waves in the air column inside the instrument. The pitch is controlled by the opening/closing of the six finger holes and the single key along the instrument, in effect changing the length of the active air column. The full length of the air column vibrating when covering all six holes and keeping the key closed produces the root of

FIGURE 8.9
The neck of a cecilium, with buttons in rows mimicking strings. Minor seconds between buttons vertically, and perfect fourths horizontally. Photo by Tellef Kvifte. Courtesy of Ringve Museum.

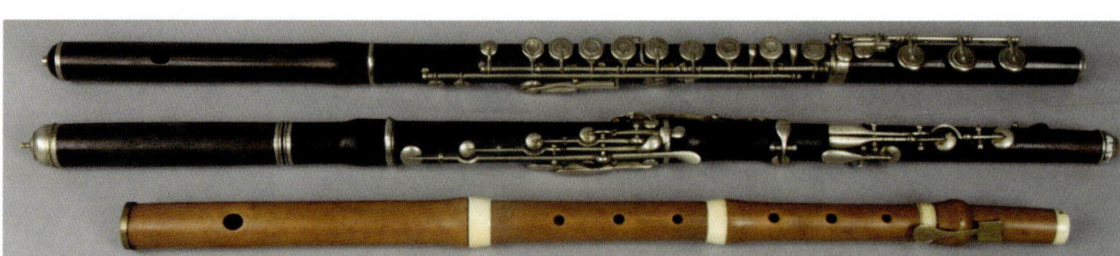

FIGURE 8.10
Three traversoes from the Ringve collection: a one-key traverso (bottom); a thirteen-key traverso (middle), basically adding a number of keys to the baroque traverso and a Boehm model (top) that has undergone a more fundamental revision of both the acoustic and mechanical properties, giving the instrument more power. Photo by Tellef Kvifte. Courtesy of Ringve Museum.

a major scale[4] that can be played by uncovering the holes one by one, and finally closing all again and blowing harder to get a note one octave above the starting note. The index, long, and ring finger of the left hand normally control the three upper holes, while the corresponding fingers of the right hand control the lower three holes. The remainder of the twelve semitones of the octave can be normally be played by various so-called fork fingerings, combinations of open and closed holes where not all the open holes are consecutive holes from the bottom (like the C natural in Figure 8.11). Such fingerings produce notes of a tone quality that differs from the notes produced by the "normal" fingerings; they may be slightly out of tune, and they are generally awkward to play in fast tempi.

The differences between the early traverso and the other orchestral woodwinds concerns the sound-producing mechanisms, the naming of the fundamental note and, in the case of the clarinet, in the interval of overblowing. However, the basic principle of a scale produced by uncovering hole-by-hole from the six-finger stop is common to all. This is a very convenient arrangement for music that sticks to the particular scale starting at the bottom note, and it is increasingly inconvenient as the music moves away from this scale and fork fingerings have to be used.

To overcome such problems, the woodwinds were furnished with a number of keys that operated on extra holes and that allowed the performer to use more fingers than the six used for the basic holes. The number of keys was generally quite small until the beginning of the nineteenth century, and roughly during the same period of time as that of the development of the free-reed instruments, new models of all the woodwinds with a larger number of keys were introduced. The aim was primarily to allow easy fingering and equal tonal quality for all twelve notes in the octave (Figure 8.10, middle and top).

The saxophone (Figure 8.12)—an instrument invented later in the century—continues the sophistication of these user interfaces and of the large amount of pads, keys, and mechanical details that were used in them. Notice also that even if the instrument is constructed with the aim of easy access of all keys and all twelve notes in the octave, it still builds on the basic principles of the traverso pitch interface.

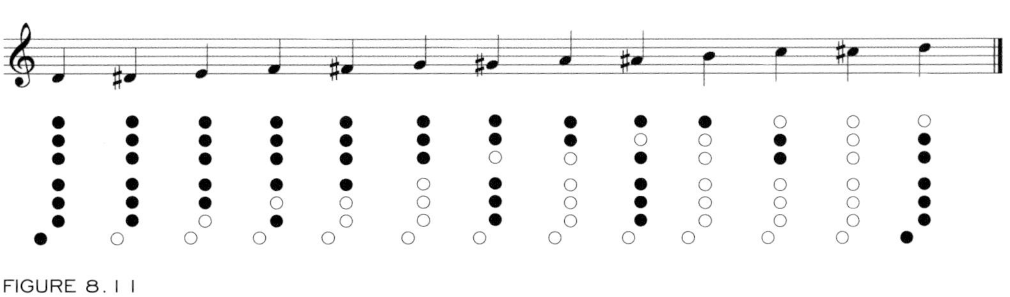

FIGURE 8.11
Traverso fingering. Black circles indicate closed holes. The circle at the bottom indicates the single key on the traverso.

FIGURE 8.12
Part of the keywork of an Adolphe Sax soprano saxophone. Notice the six larger ivory pads, corresponding to the six basic finger holes on the traverso. Photo by Tellef Kvifte. Courtesy of Ringve Museum.

Twentieth-Century Interfaces

Turning to the twentieth century, we find a different focus in the development of user interfaces. It is tempting to start with the Theremin (Figure 8.13) as a complete contrast. Instead of a large number of buttons or keywork, the user interface is stripped down to two metal bars, in the form of an "antenna" on the top and a loop at the left of the instrument. The instrument is controlled by moving the hands relative to these bars: pitch is controlled by the distance between the right hand and the antenna, and loudness by the distance between the left hand and the loop. In contrast to the free reeds as well the woodwinds, there is no representation of discrete scale steps in the user interface, and pitch can be varied continuously. The contrast to the many buttons and keys of the nineteenth-century interfaces could hardly have been more striking.

While the Theremin in many ways may be seen as both a starting point and a "typical" twentieth-century electrophone, it is quite atypical in its almost minimalistic user interface. Quite complex user interfaces seem to be a rule rather than an exception in the

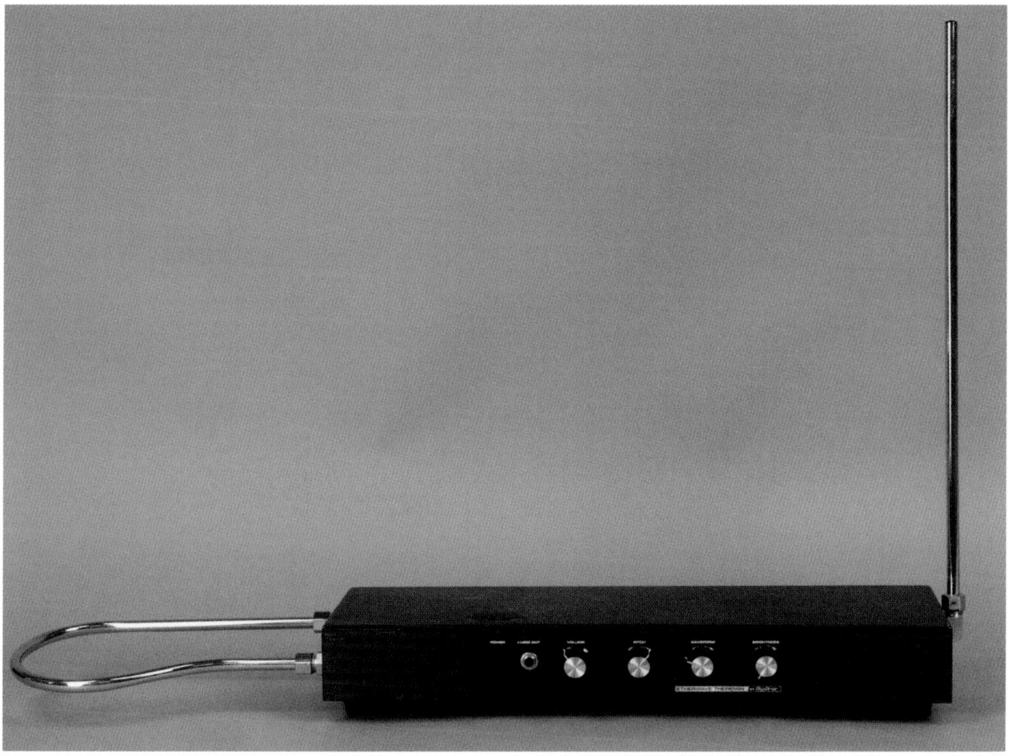

FIGURE 8.13
Theremin. Photo by Tellef Kvifte. Courtesy of Ringve Museum.

later development of instruments, but, as the Theremin forewarns us, the interest is no longer primarily the control and organization of scale steps.

Early Synthesizers

When I worked in NRK (Norwegian Broadcasting Corporation) in the beginning of the 1970s, the head of the music section, Gunnar Sønstevold, had started to build a small studio for electronic music. Besides a mixing console and some tape machines, there were also a VCS3 synthesizer and a Subharchord (Figure 8.14), a machine developed in Eastern Germany and produced in a very small quantity. The Subharchord looked in a way more like an "instrument" to me than the VCS3, no doubt because of the familiar piano keyboard.

Nevertheless—or perhaps just for that reason—it was the VCS3 that drew my attention, and I spent quite some time with it. The Subharchord, on the other hand, was not much in use; in fact, I cannot recall ever hearing a sound from it during the few years

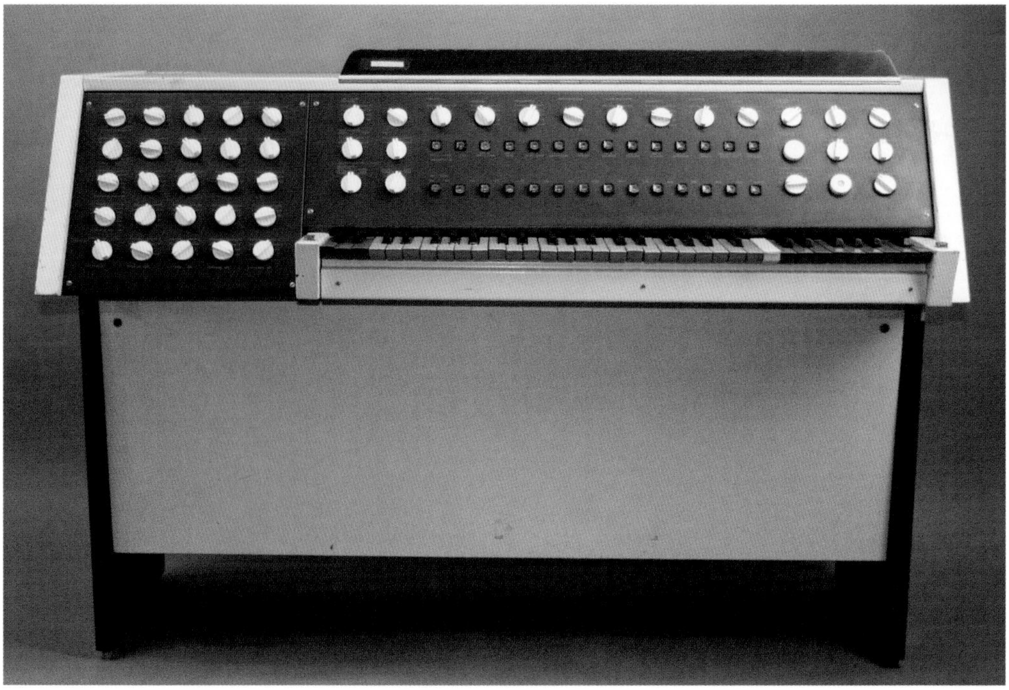

FIGURE 8.14
The Subharchord, once in use at the Norwegian Broadcasting Corporation. Courtesy of Ringve Museum.

it was available in the studio. It is now part of the collection of the Norwegian Museum of Science and Technology, and on display at the Ringve Museum.

The first thought I had when I saw the VCS3 for the first time was "How is this thing operated?" Its controls were all different from what I knew of instruments at that time, and, most important, I could see no keyboard-like control organ to play the different pitches of a scale. As it turned out, there *was* no such control, but lots of others of different kinds. Later I learned that a keyboard *could* be added to the unit, as a special option. In contrast to other instruments I had seen, most of the controls are knobs to be twisted and turned, and not keys or buttons to be pressed,[5] and what is said in the manual is quite pertinent:

> The VCS3 is not "played" like a conventional musical instrument, but it is all the same capable of a far greater range of sounds than any one musical instrument, and since its controls are all continuously variable the varieties of sounds obtainable are literally endless.[6]

Hardly an exaggeration, though the sounds are easily recognizable as "electronic" and—in many cases—possible for an expert to identify as coming from a VCS3. Nevertheless, and even if the VCS3 is quite simple in construction compared with later electronic instruments, it is a good example of important characteristics of the new instruments. The VCS3 is built of a small number of modules (called "devices" in the user manual for the VCS3) that can be connected in different ways.[7]

When the VCS3 is turned on, the instrument is silent, and, unlike "normal" instruments, has to be "set up," "configured," or *patched* to be able to produce any sound at all. Of course, all instruments have to be prepared in some way, like opening the lid of a piano, wetting and mounting the reed on a clarinet, tuning the fiddle; but this setup is of a different kind, as it will not only prepare the instrument for normal operation but also will *define* both the actual sound and how the controls of the instrument will affect the sound.

This setup is done in two stages, usually with some iterations. First, a number of modules are connected by putting pins in different locations on the patchboard (Figure 8.15). The general idea is that a pin in a given square shown in Figure 8.15 will connect the output of the module named at the left of the row to the input of the module at the top of the column.

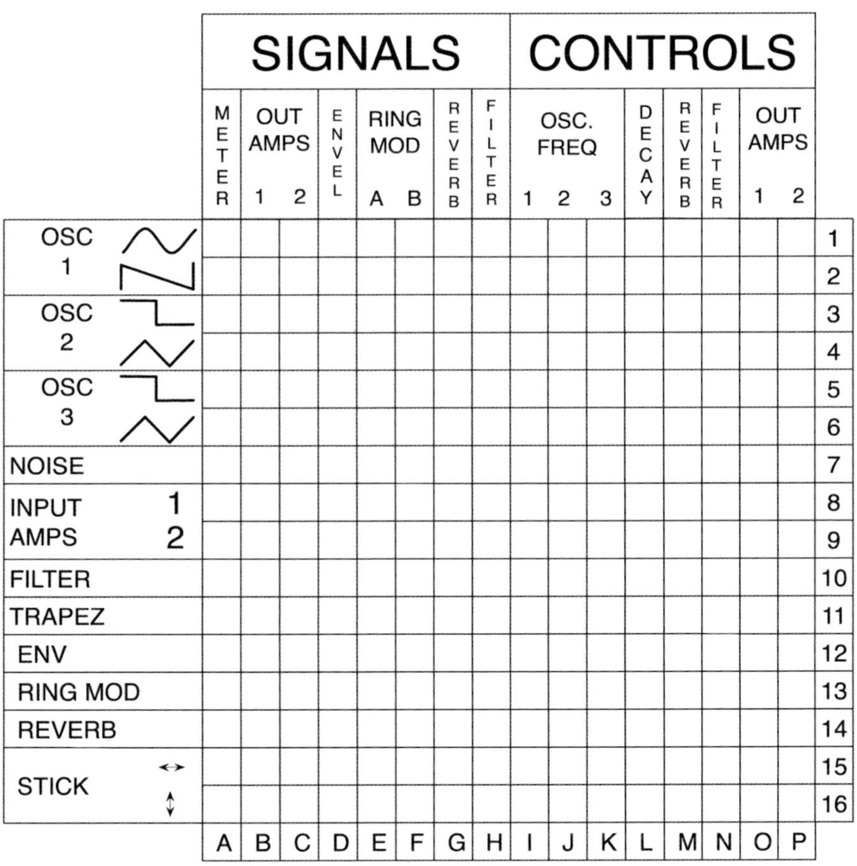

FIGURE 8.15
The patchboard layout of the VCS3.

Note that the inputs at the top are grouped into "signals" and "controls." The implication is that not only can a sound signal be passed from one module to another but also the signal from one module can be used to *control* how another module works, as explained in the user manual:

> Traditionally designed source or treatment devices have only manual means of varying their principal parameters, whatever they may be (and this depends on the function of the device of course). The circuits in the VCS3, on the other hand, have as well as manual control the possibility of voltage control. By this we mean that by applying a voltage to a special input on the device its main parameter can be automatically varied . . .
>
> For example, the control input of an oscillator will vary the frequency, that of an amplifier the gain, etc. What gives the VCS3 its great versatility is that one device can be made to act on the control input of another, and whole chains of interdependent events can be built in this way.[8]

The Big Modulars

The VCS3 is a rather small box compared with other modular synthesizers of the period, most notably the Moog modular that was used by several important pop artists, and also the lesser-known Roland system 700 that was produced in a much smaller quantity than the Moog. One of the Roland systems was bought by the Department of Musicology at the University of Trondheim in the second half of the 1970s for use in education; later it was moved to the Ringve Museum, where it is exhibited today (Figure 8.16).

The basic principles and types of module are the same as in the VCS3, but there are more modules of each kind, and the connections are made by stretching patch cords between the modules (as on the Moog) rather than by putting pins in a matrix (as on the VCS3). The number of sliders and knobs is great; far greater than the number of buttons or keys on the nineteenth-century interfaces, and the operation is also made more complicated by the fact that the exact effect of the individual knob and slider on the final sound depends on the patching—how the modules are connected. A great deal of the control possibilities are concerned with timbral qualities, and the controls for pitch are more concerned with gradual variation than with the control of scales and chords.

Other Examples

It may seem to be a long way from the VCS3 to the hardware and software synthesizers and samplers that are used in contemporary music production. However, many of the basic characteristics are the same. The screenshot of the software instrument Zebra (Figure 8.17) is one example. The figure shows just one of several screens of the instrument, each with different controls and information. On this screen, we can see a number of

FIGURE 8.16
The Roland 700 Modular. Photo by Tellef Kvifte. Courtesy of Ringve Museum.

FIGURE 8.17
Screen shot of part of the user interface of the software synthesizer Zebra. Courtesy of Urs Heckman.

modules, many of the same kind as in a VCS3 (oscillators, filters, envelope generators, etc.) and a number of others as well. Each has a number of continuously variable controls and a few stepwise. The modules are stacked on each side of a central window where, as in the VCS3, the modules can be connected in different ways. Also, similar to the VCS3, most controls affect timbral qualities rather than stepwise pitch.

Compared with the VCS3, the possible number of modules to be used is much bigger, and, important from a practical point of view, any setup can be stored for later instant recall, regardless of the complexity of the patch. On the VCS3, one had to manually write down on paper all the settings and manually reset all the controls to re-create a patch on a later occasion. Also, even if one took great care in the description of the patch, one could not be sure that a reconstructed VCS3 patch would sound the same as the original, because the modules were inherently unstable. The contemporary counterparts do not present these problems.

Discussions

Definitions of Instruments and Mapping

To sum up so far, we have seen interfaces from two different periods of instrument evolution; the development of free-reed instruments in the nineteenth century and of synthesizers in the twentieth. In both cases, a large number of new interfaces were developed, based on new methods of sound production. To describe differences and continuities in more detail, it is useful to make some remarks on the central concept of "musical instrument." There are many ways to define what kind of object a "musical instrument" is, and we shall not go into this in any great detail here.

The perspective used in this chapter builds on the work of the Polish musicologist Ludwik Bielawski. He describes musical instruments as *transformers* that transform the gestures of a musician into musical gestures.[9] This way of defining musical instruments emphasizes control organs and user interfaces, the parts of the instrument that musicians can affect in different ways to make the instrument produce a range of sounds we can perceive as music; the parts that we have been concerned with so far. But the central idea in Bielawski's definition is not the user interface as such, or the resulting sound. It is rather the *relationship* between the gestures of movement and the resulting musical gestures, as indicated in Figure 8.18.

The action of *finding* a certain key on a keyboard is, in Bielawski's diagram, a *space* movement. Finding a specific place in space, each key and each place is connected to a specific pitch. *Striking* the key—a dynamic action in Bielawski's terminology—is connected to the loudness of the sound. This relationship between playing actions and musical sound is also known as *mapping*. Each key on a piano (or on a marimba, or other similar instrument) is *mapped* to a specific pitch. The force of striking the key (or the drumhead, or the marimba key) is mapped to the loudness of the sound. In many ways, the mapping may be regarded as more important to the identity of an instrument than

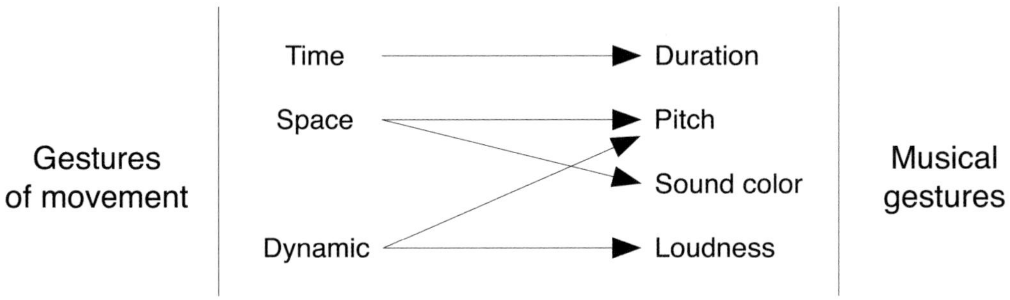

FIGURE 8.18
Bielawski's diagram of connections between gestures of movement, or playing actions, and music.

the user interface or the sound as such. Consider, for example, the user interface of keyboards, used on such diverse instruments as church organs, pianos, accordions, and synthesizers; or consider a flutelike sound that may be produced by such different instruments as flutes, organs, or synthesizers.

While mappings are specific to each instrument, there are—at least in acoustical instruments—some broad generalizations to be made, as indicated in Bielawski's diagram. For instance, loudness is usually controlled by a "dynamic action," meaning in practice that we, as listeners and spectators of musicians playing an instrument, will expect the music to get louder when the performer uses larger and faster movements. On many instruments, pitches are laid out in some systematic fashion in space, so we expect pitches to get higher when a pianist (or a marimba or organ player) reaches out to the right, or when a cellist reaches further down on the fingerboard. The patterns of such regularities are taken for granted when we watch and listen to a musical genre we know and are shared by musicians and listeners. It can similarly be said that mappings are important for the identity of an instrument, as the specific "choreography" of the playing movements of the performer is characteristic of each kind of instrument.

Differences in Mapping

With Bielawski's general scheme and the concept of mapping, we can have a closer look at the relationship between the acoustic instruments of the nineteenth century and the electronic ones of recent times.

One of the most obvious differences is that *mappings can be changed* on the new instruments. How a certain control action influences the sound is programmable, and any part of a user interface can be mapped onto any of the available controls of the sound-producing mechanism. How one sets up the VCS3 will determine the effect you get (if any) when turning a certain knob on the instrument. On a modern synthesizer, one can set up the keyboard so that the velocity is mapped to loudness (as is usual) or a filter setting, so that harder hits give brighter sound, or to the envelope so that harder hits give longer sound, and so forth.

What is more, it is also possible to set up mappings that are experienced as fundamentally "unnatural," as Wendy Carlos found when trying a VCS3: "It also has a so-called touch-sensitive keyboard which has to be tried to be believed, it's that awful . . . (and the one I tried worked backwards: softer touch = louder sounds!)."[10]

One further striking contrast that calls for some discussion and speculation is the obvious difference in what the respective user interfaces control. In the nineteenth century, the control of pitches—scale steps—is the central theme. This is evident not only from the wide variety of free-reed instrument interfaces that evolved; the same tendency is also obvious in the many new designs of woodwinds that during this period focused on bringing all twelve notes of the octave within comfortable reach of the musician, both in terms of convenient fingerings as well as with even tonal quality.

In the twentieth century, the focus is on expressive and timbral qualities rather than on pitch control. The many controls and the large number of possible patches on the VCS3 are used first and foremost to control and vary timbral qualities. The keyboard, to control pitches in a more ordinary fashion, was just an optional add-on.

Connections

The development of instrument interfaces is not an activity done in isolation; like any technological change, it is connected to economic, cultural, and aesthetic processes in the society at large. In the concluding part of this study, we shall try to connect the development described so far to the dominating tools for distribution and production of music in the two periods, implying that there is a mutual influence among the technologies of instrument design, of music production, and of music distribution (Figure 8.19). In short, the argument is that standard music notation dominated in the nineteenth century, and sound recordings took over in the twentieth; and that the increasing complexity of the control organs for pitch in the nineteenth century is connected to the central position of musical notation. In the twentieth century, the increasing complexity of the control organs for timbre is connected to the central position of sound recording and transmission.

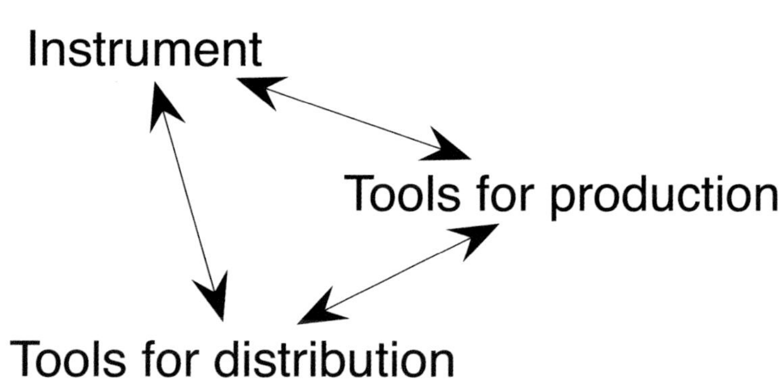

FIGURE 8.19
Three interconnected areas of technology.

Nevertheless, it is not always possible to separate the three concepts of instruments, tools for music production, and tools for distribution of music. In the nineteenth century, standard music notation was both a medium for the dissemination of music, and at the same time, an important tool for composers in the production of musical works, as well as for musicians in the production of the actual sound. Likewise, the sound reproduction media of the twentieth century were increasingly used for the production of the music and not only as passive recording devices and tools for the distribution of music.

Also, the distinction between instrument and distribution technologies is not always clear. In the Pianola and various music machines used for the reproduction of music, actual acoustic instruments are controlled by mechanical means to re-create music that has been "programmed" on perforated paper, pegged cylinders, or some other similar contraptions.

The close relationship between these three areas shows up in a number of ways. Starting with the nineteenth century, one possible observation is that the dominating distribution and production tool of standard notation is very well suited to the description of patterns of pitches in the form of melodies and chords (and also to duration classes like whole note, half note, quarter note . . .). But standard notation is not very well suited to describing *inflections* of pitches, nuances of rhythmic timing, and even less suited to describing timbre and variations in all these qualities, which are, in other words, the musical attributes usually perceived as the domain of the performer rather than the composer. This corresponds well to the dominance of pitch-control interfaces on the nineteenth-century instruments described above, where scale steps are favored over intonation and pitch inflections; not to speak of timbral variety. Standard notation as a tool of production also corresponds well to the general focus on development of pitch patterns in the form of melodies and, even more obvious in the art music of the century, the development of complex, chromatically flavored harmony. While timbral qualities certainly were pursued to some extent, especially in the great symphonic works of the nineteenth century, some scholars nevertheless regard this period as "pitch-dominated"; a tendency that continued long into the twentieth century. Cornelia Fales observes, when commenting on a series of recordings from the 1950s of a Burundi tradition, when the ethnomusicologist Alan Merriam in his placement of the microphone favored the pitches of the accompanying zither rather than, according to Fales, the musically all-important timbral qualities of the voice in this genre:

> Merriam's recordings of the music betray the subtle bias of what has come to be called "pitch-centrism" or "timbre deafness," a perceptual proclivity on the part of western listeners, including ethnomusicologists, to focus on melody in music where the dominant parameter is timbre.[11]

The music theory of the period is to a very large part concerned precisely with the ordering of pitches in scales and chords, and the concepts of music theory form a common language that can be used to connect the three areas: names of notes, for instance, are used to describe both the dots on the music paper and the keys, buttons, and finger holes on the instrument. These connections are frequently made explicit, like in Figure 8.2 and in fingering charts as in Figure 8.11. The relation between standard notation and the piano keyboard is also an example of the close connection between instrument construction and distribution technology, with the common prominence of the C major scale through "white keys only" and "no accidentals" respectively.

The music distribution technology of sound recordings is quite different. Here, sound (both in the literal sense and when understood as "sound" as a musical distinguishing element) and timbral quality can be communicated in great detail, while precise pitches may be harder to get at, as any music student who has tried to decipher, say, the chord voicings of a jazz pianist on a recording can testify. The identity of the musicians, composers, and arrangers may admittedly be printed on the sleeve of a recording, but the *recorded sound* as such does not spell out their names. Nevertheless, the *sound* of a musician can easily be recognized, at least by the regular audience of a given style; and in many genres, musicians strive for a distinctive sound that can be easily recognized on a recording. This is also what Katz argues when he explains why the violin vibrato in classical performances increased significantly after the introduction of recording technology:

> I would suggest that a constant and strong vibrato became increasingly useful for concert violinists who regularly made recordings, and it did so in three ways. First, it helped accommodate the distinctive and often limited receptivity of early recording equipment. Second, it could obscure imperfect intonation, which is more noticeable on record than in a live setting. And third, it could offer a great sense of the performer's presence on record, conveying to unseeing listeners what body language and facial expressions would have communicated in concert.[12]

The advent of sound recordings, and their increasing adoption in the twentieth century, also gave *musicians* a tool for further dissemination of their particular musical expertise, namely the expressive qualities of *performance* of music, such as timing, intonation, and the shaping of timbral qualities of the performance. Such qualities were hardly possible to communicate through standard notation and were therefore out of the control of composers who relied on notation as their main tool for production and distribution. The emerging recording technology was, at the outset, however, not a tool for shaping and working with these expressive qualities to any great extent. It was suited to, and thought of, primarily as a reproductive tool. Therefore, it is tempting to view the developments in the twentieth century in two overlapping stages, where the first is dominated by the rise of performers through dissemination of their performances through recordings, while the second, that for all practical purposes may be regarded as starting with the rise of the magnetic tape recorder, is dominated by the emergence of new tools for working with those expressive qualities that until now were the domain of performers. Among these tools are the twentieth-century instruments that we have discussed, and many others. One of the most important is multitrack recording that made it possible to isolate single instruments in recordings; and then there came a great array of sound-processing equipment that could be applied to the recorded sounds, including reverbs, filters, flangers, and compressors. Recording technology had evolved from a reproductive to a true productive tool.

Resisting New Technology

In the two periods we have covered, the new technologies described perhaps inevitably met with resistance in different ways, and arguments made against them can also give us some insights. New possibilities always come at the expense of something, and the evenness of tone and easy access of all pitches on the new woodwind interfaces were no exception. Grove's online observes that "some oboists thought that too many keys could damage tone quality," showing that there were different views on what constituted "tone quality," and that "evenness of tone" was not a generally accepted aesthetic view. There were, indeed, more sacrifices resulting from the new keywork mechanisms:

> The increase in the number of keys—which went hand in hand with the enlargement of the sound holes to match the acoustic requirements—without a doubt decreased the contact between the fingers of the player and the sound hole. This had far-reaching consequences. The possibility of creating a vibrato by simply moving the fingers over the sound holes, which until that moment in history had been essential to the realization of sound and articulation, was ruled out. If the musician did not want to part with the vibrato altogether he had to replace it with a deliberately produced lip vibrato. However, the range of nuance created was not nearly as extensive. Forked fingerings became expendable, again a fact that brought about drastic changes. What was regarded an important boon by Theobald Boehm and others, though, was considered a loss by many a musician. In fact, they no longer had full control over which finger position they deemed best to correct out-of-tune sounds (choices where technical considerations were as decisive as preferences for certain finger combinations and the particular acoustic idiosyncrasies of each instrument, etc.).[13]

It should be noted that neither tone quality nor vibrato are possible to prescribe or describe with any real precision in standard notation, and were therefore, as argued, out of control for the great composers of the era. What the performers had to give up with the new interfaces was therefore part of their exclusive control over the musical performance. But as the composers took the ready availability of all pitches and pitch combinations more or less for granted, the resistance against these interfaces subsided significantly during the century, at least among art music performers, who were increasingly dependent on the new instruments to be able to play the music as required by the composers.

In genres where diatonic scales in keys with few accidentals continued to dominate and standard notation was little used, the situation was different, and the older systems continued to prevail. Flutes used in Irish traditional music are to this day almost never of the "modern" Boehm type but are of some older type with no rings, allowing for direct contact between the fingers and the six main holes.

Turning to the twentieth century, resistance to electric and electronic instruments took many forms (some as described by Frode Weium elsewhere in this volume),[14] frequently in the form of criticism of the "machine-like qualities" perceived in the new instruments—not too surprising, given the new level of technological complexity. But also the change from the focus on complex pitch patterns of nineteenth-century orchestral scores to the meticulous timbral shaping of late twentieth-century[15] record productions made for some interesting situations. Expectedly, composers who were used to standard notation as a productive tool, and, not least as the medium that to a large extent defined what "true music" was, were hesitant. Much of their cultural capital was invested in this tool, and, it can be added, some economic capital, as many of the copyright practices were built around notation, a "musical work" being for a long time identified with the written score. When tools that gave performers some power as well—namely the recording technology and the sound-shaping tools that followed in its wake—were being used as productive tools, they were to a large extent defined as "low culture" or "commercial" by the art music establishment. The avant-garde composers, however, needed some finer distinctions. As they to a large extent relied on the same technology as the "commercial" music, distinctions made on the basis of use rather than technology were sometimes more convenient.

I stumbled upon one such distinction in the late 1970s. At that time, the Norwegian Broadcasting Corporation ran a small electronic studio together with the Composers' Guild, built around a Buchla synthesizer—*the* art music synthesizer of the period, and quite distinct from the Moog synthesizers that were perceived as commercial products. At one point, I wanted to use the studio to produce some sounds for use in a television program. I came to the studio with a recording of some music made by the band taking part in the program, and we had an idea to process this particular piece of music in the Buchla machine and incorporate the passage in the otherwise live performance of the band. I was not allowed to do this, because the use of prerecorded sound was defined as "sampling," and there was a decision from the board of the studio that sampling, being a commercial activity, was not allowed in the studio.[16]

This was not an isolated event—the technique of sampling[17] has continued to be attacked up to this day.[18] Stories like this may be funny now, but they show clearly how technologies are fundamentally tied into their contemporary cultural, artistic, and economic systems.

Conclusion

In the introduction, I asked, "In what ways are electronic instruments different from acoustic instruments?" The way I have argued, there is no simple answer to that question. I have tried to show that instrumental practice is deeply embedded in the musical culture at large, implicating that the important units of study are not "instruments" as such, but some larger units incorporating also music theory and practice, as observable in music production and distribution.

Or, in other words, the introduction of the free reeds is not only about the novelty of a new sound-producing device. It is also part of a musical culture with strong focus on pitch patterns (melody, harmony) at the expense of timbral and rhythmic qualities; a focus that both led to experiments and stylistic developments where all twelve pitch classes of the octave had to be easily accessible for the performer, as in the music of Arnold Schoenberg and the later serial music on the one hand, as well as to novel ways of organizing user interfaces for musical styles of a predominantly diatonic character on the other.

Likewise, the introduction of electronic instruments is not primarily "about" electricity or "digital technology." As I have argued, it is also about aesthetic experiments in the control of musical performance and experiments in the control and production of timbral qualities, a development we have not yet seen the end of, by a long way. As a final point, one should observe that acoustic instruments are also part of this development, with increasingly sophisticated techniques to increase their timbral palettes being invented by contemporary performers.

Notes

1. Willis and Company, *German Æolian Tutor*, London, 1830, 14.
2. The "Accordion" entry in *Grove Music Online* estimates the number of varieties in use as somewhere between forty and fifty-five.
3. Charles Wheatstone, "Concertinas and Other Musical Instruments," England 1844, 3.
4. This changed to minor scale on several new key systems.
5. This was also the case of the Subharchord, with the exception of the keyboard.
6. "VCS3: The Putney Compact Electronic Music Studio User's Manual," Electronic Music Studios (London) Ltd., 1970, 5.
7. The modules include three oscillators that produce pitched sounds, and a noise generator. Further, some units were used to modify the sounds: filter, reverb, ring-modulator, and output amplifiers.
8. "VCS3: The Putney Compact Electronic Music Studio User's Manual," 5.
9. Ludwik Bielawski, "Instrumentalmusik als Transformation der menschlichen Bewegung. Mensch–Instrument–Musik," *Studia Instrumentorum Musicae Popularis VI* (1979): 27–32.
10. Walter Carlos, "On Synthesizers," in *The Last Whole Earth Catalog* (Harmondsworth, UK: Penguin Books, 1971). (Needless to say, the keyboard can also be set up in "normal fashion.")
11. Cornelia Fales, "The Paradox of Timbre," *Ethnomusicology* 46(1) (2002): 56–95.
12. Mark Katz, *Capturing Sound: How Technology Has Changed Music* (Berkeley: University of California Press, 2005), 93.

13. Christian Ahrens and Irene Zedlacher, "Technological Innovations in Nineteenth-Century Instrument Making and Their Consequences," *Musical Quarterly* 80(2) (1996): 332–33.

14. Frode Weium, "Technology and Authenticity: The Reception of the Hammond Organ in Norway," in this volume.

15. . . . and early twenty-first century, for that matter.

16. This is how I remember it—I have been unable to find written evidence of this today, so I can't say for sure that there actually was such a decision, nor what wording it might have used.

17. "Sampling" has several meanings, among them the use of recorded sounds as a basis for instrumental sounds or for further manipulation in new recordings/compositions; see, for example, Tellef Kvifte, "Digital Sampling and Analogue Aesthetics," in *Aesthetics at Work*, ed. Arne Melberg, 105–28 (Oslo: Unipub, 2007), for a discussion.

18. See, for example, Andrew Goodwin, "Sample and Hold: Pop Music in the Age of Digital Reproduction," in *On Record: Rock, Pop and the Written Word*, ed. Simon Frith and Andrew Goodwin (London: Routledge, 1990). See also Tara Rodgers, "On the Process and Aesthetics of Sampling in Electronic Music Production," *Organised Sound* 8(3) (2003): 313–20, for different and opposing views.

Bibliography

Ahrens, Christian, and Irene Zedlacher. "Technological Innovations in Nineteenth-Century Instrument Making and Their Consequences." *Musical Quarterly* 80(2) (1996): 332–40.

Bielawski, Ludwik. "Instrumentalmusik als Transformation der menschlichen Bewegung. Mensch–Instrument–Musik." *Studia Instrumentorum Musicae Popularis VI* (1979): 27–32.

Carlos, Walter. "On Synthesizers." In *The Last Whole Earth Catalog*. Harmondsworth, UK: Penguin Books, 1971.

Fales, Cornelia. "The Paradox of Timbre." *Ethnomusicology* 46(1) (2002): 56–95.

Goodwin, Andrew. "Sample and Hold: Pop Music in the Age of Digital Reproduction." In *On Record: Rock, Pop and the Written Word*, ed. Simon Frith and Andrew Goodwin. London: Routledge, 1990.

Katz, Mark. *Capturing Sound: How Technology Has Changed Music*. Berkeley: University of California Press, 2005.

Kvifte, Tellef. "Digital Sampling and Analogue Aesthetics." In *Aesthetics At Work*, ed. Arne Melberg, 105–28. Oslo: Unipub, 2007.

Rodgers, Tara. "On the Process and Aesthetics of Sampling in Electronic Music Production." *Organised Sound* 8(3) (2003): 313–20.

"VCS3: The Putney Compact Electronic Music Studio User's Manual," Electronic Music Studios (London) Ltd., 1970.

Weium, Frode. "Technology and Authenticity: The Reception of the Hammond Organ in Norway." This volume.

Wheatstone, Charles. "Concertinas and Other Musical Instruments." England (1844): 3.

Willis and Company. *German Æolian Tutor*. London, 1830.

CHAPTER 9

Austrian Pioneers of Electronic Musical Instruments

Peter Donhauser

Introduction

The Vienna Museum of Technology was founded in 1909 and opened in 1918. The main part of the collections goes back to the nineteenth century, but some artifacts are much older (an astronomical clock and a small organ of the late sixteenth century are two fine examples of these).

The museum owns a particular collection of musical instruments. This is due to its tradition as a museum of craft and industry, for in nineteenth-century Vienna instrument making was a major branch of industry. The core of the collection goes back to that time. Nor is it just any random collection of musical instruments: the focus is on Viennese fortepianos together with self-playing and electronic instruments, special attention being paid to such aspects as their manufacture and function.

A new section dedicated to electronic instruments and synthesizers was established recently, including pioneering Austrian constructions (Figure 9.1). Until a few years ago, little was known about the development of electronic instruments in Austria. A research project has provided the necessary background information for documenting the collection. The present survey outlines the development of these artifacts in Austria between 1920 and the late 1950s, including the most recent findings.

Before discussing these developments in detail, it is necessary to make some remarks on nomenclature. The present-day differentiation between *electrical* and *electronic* (respectively without and with the usage of active components, including valves or transistors) was unknown in the time we are discussing. Until the start of *electronic* music in today's meaning in the 1950s, only the term *electrical* was used. Jörg Mager (one of the

FIGURE 9.1
The Austrian pioneer synthesizers designed at the Vienna University: AKA 2000 and Akaphon. Courtesy of Peter Donhauser.

protagonists of these new instruments in Germany) tried to promote the use of the term *Elektrophon*, but without response. Following current usage, however, this chapter has standardized on the term "electronic."

The European Context of Electronic Instruments of the 1920s and 1930s

The roots of electronic music go back to the late nineteenth century, when the American patent attorney Thaddeus Cahill built his Telharmonium, a two-hundred-ton instrument using AC generators similar to those used in power stations. While interest in the instrument soon waned due to its technical and functional complexity, the idea was to inspire many others over the following decades.

The Aetherphone, invented by the Russian engineer Lev Termen, was to have a more lasting influence. Termen first presented his instrument, which is based on the heterodyne principle, at the All-Soviet Electrical Engineering Congress of 1921. Performances by Termen in Frankfurt, Berlin, and London in 1927 (from that time he called himself *Thèremin* according to the Gallic roots of the name), and later in the United States, not only fascinated listeners but also engineers. Reproductions and copies were built all over the Western hemisphere. The Aetherphone, or Theremin, is the only electronic musical instrument of that time still used today.

The first people to experiment with electronic musical instruments in Germany were Jörg Mager (numerous designs, including the Sphaerophon and the Partiturophon), Bruno Helberger (Hellertion), and Friedrich Trautwein and Oskar Sala (Trautonium). They were soon to be followed by Oskar Vierling (Grosston organ and Elektrochord) and Harald Bode (Melodium). The Neo-Bechstein of 1931 was the first electric piano to be produced in a small quantity. The instruments designed by these pioneers may have remained marginal, but the crucial ideas behind them certainly have not. Generators and interfaces, such as the ribbon controller and the filters used in the Trautonium and the Hellertion, are still in use today.[1]

Similar developments took place in France (Maurice Martenot's Ondes musicales and Armand Givelet's electronic organ) and in Russia (for example, Evgeny Sholpo's Variophone and Georgy Rimsky-Korsakov's Emiriton). Some activity can also be traced in other European countries (for example, in the Netherlands, where the Trautonium was presented by Oskar Sala; and in Norway, where several attempts were made to introduce electronic instruments.[2])

Judgment of the aesthetic qualities of these instruments based upon an evaluation of their sounds is difficult. As original examples of the instruments are mostly missing or in a nonworking condition, we are reliant upon contemporary recordings. The Theremin, Trautonium, Neo-Bechstein, and Ondes musicales (known as Ondes Martenot) are well represented in early recordings. Sholpo's recordings were discovered some years ago. Relevant recordings of the *Berlin Radio Show* of 1932 are preserved at the Folkwang Hochschule, Essen. Recently a recording of a Jörg Mager instrument was discovered, which was originally thought lost.[3] At the moment no recordings of Austrian instruments are known, with the exception of one of them (the Heliophon), which was preserved on a private tape recording and on two films by the Austrian Broadcasting Corporation.

The Situation in Austria

Compared to Germany or Russia, for example, the development of electronic musical instruments in Austria took a different course. There was no political interest in this new technology, as there was with the Trautonium and the Hellertion in Germany or the Theremin in Russia.[4] This might be explained by the unstable situation after World War I ending in a civil war, an authoritarian regime, and finally the Anschluss annexation of Austria to Nazi Germany. Political parties at that time showed no interest in electronic instruments as a means to promote their aims; they were too underdeveloped, and the protagonists were not as politically organized as Trautwein or Helberger were. The only consequence of the political situation was the end of Emerich M. Spielmann's and Robert Pollak-Rudin's work (see below).

Neither did the art universities in Austria show any interest in electronic musical instruments. This is completely different from the situation in Berlin, where the Hochschule der Künste and the Heinrich Hertz-Institut played an active role in the development and promotion of the instruments. As the Austrian musical culture was generally rather conservative, there were no major initiatives coming from artists or even from the media. Thus, the development described below is characterized by gaps and discontinuities.

"The spirit of research and old-Viennese craftmanship"[5] was how the *Neue Freie Presse* hailed Rudolf Stelzhammer's Magneton on August 14, 1930. "Ground-breaking scientific pathways were embarked on for the first time"[6] the *Neues Wiener Extrablatt* commented on the Chromatophon, designed by Anatol Vietinghoff-Scheel, on December 10, 1929.

While the development of electronic instruments and the use of electricity in instrument making in Austria during the 1920s and 1930s does not match other European countries in extent and dynamics, it still produced some remarkable results. Among them, there is an optical-sound instrument, Emerich Spielmann's Superpiano of 1929, using optical tone discs, and another based on magnetic effects, the Magneton referred to by the *Neue Freie Presse*, created by Stelzhammer and Wilhelm Lenk in 1930. A documentation of the above-mentioned research project has recently been published.[7] But as research is ongoing, new findings—for example, on the role of Ernst Werndl (a nephew of the Austrian industrialist Josef Werndl), about which very little was known until recently—show the instruments in an entirely new light.

A closer relationship of the development of electronic musical instruments between Germany and Austria cannot be found. The protagonists knew of each other (for example, we find the Superpiano in Peter Lertes's book *Elektrische Musik* of 1933; Spielmann and Werndl mentioned the German developments in their publications), but a closer contact did not take place.

Compared with other fields of scientific breakthrough, like the discovery of X-rays or the first wireless transmission by Guglielmo Marconi, the experiments with electronic musical instruments were delayed for thirty years. Especially in Austria, with a long and conservative tradition in instrument manufacture, the enthusiasm for new musical trends was kept within limits. On the other hand, electrical engineering also has a long-lasting history in Vienna. Both traditions were strictly separated except for incidental collaborations. So the cooperation of a technician like Lenk and an instrument manufacturer like Stelzhammer could be seen as a "wedding" between tradition and modernity, but in truth it was an attempt to overcome economic shortage.

All the Austrian instruments of those times remained prototypes, none of them being produced in quantity. Two of them are now owned by the Vienna Museum of Technology, but neither is in playable condition. While a full reconstruction of these instruments seems rather hopeless at present, it would certainly be desirable as it is the only way to evaluate contemporary press reports in terms of how far there is a relationship between expectations, conservative and progressive attitudes, and acoustic reality. Also, the question remains as to whether fascination with the new sounds did induce observers to overlook any potential shortcomings in the design of the instruments. To assess this, the instruments would need to be in working condition. Nevertheless, a retrospective judgment of the impression of new sounds on the auditors of the time seems to be not really significant.

Superpiano and "Thirring Piano"

The Superpiano was very likely the first fully functional optical-sound instrument designed in Europe. Its maker, Viennese architect Emerich M(oses) Spielmann, was evidently inspired by the sound-on-film process introduced not much earlier, and by the experiments of the Viennese physicist and university professor Hans Thirring, which led to the construction of an optical tape recorder called Selenophon.

Following the first ideas for the recording of sound by optical means by Arthur French St. George (London), who had a patent filed in Germany in 1883,[8] the further development began in 1888 with a "light siren" patented by Ernest Mercadier for telegraphy purposes[9] (similar to a pneumatic siren; a ray of light is intermittently disrupted by a rotating disc). This established the principle for an instrument based on sound generated by the action of light. Spielmann explicitly refers to these developments in a leaflet.[10]

A further step, which led to the presentation of a functioning light-tone instrument by Edwin Welte, but not until 1936, was a patent filed by Richard Michel, a teacher from Berlin, in 1925: "Tasteninstrument zur Erzeugung von Musik auf elektrischem Wege" ("Keyboard Instrument for the Production of Music Based on Electricity").[11] Michel's apparatus featured rotating celluloid or glass discs, on which the sounds "from pieces of music" were encoded on a medium "coated with a photographic layer." Lamps and selenium cells or rubidium-amalgam tubes translated the recorded sounds into electrical vibrations. It is not known if Spielmann was familiar with this patent. It certainly was featured in *Zeitschrift für Instrumentenbau*.[12]

In 1927, two years after Michel, Spielmann applied for a patent for an almost identical design.[13] Again, rotating discs acted as the media carrying photographically "registered" instrument sounds, "readout" by means of light bulbs and selenium cells. There were twelve discs featuring seven "tracks" (concentric circles encoded with the sound recordings), each thus creating a seven-octave instrument. The bulbs were switched on by means of a special resistor that could be adjusted by pressing a key for a soft attack. Pressing down a pedal raised the voltage of the bulbs and thus the instrument's sound volume. The twelve celluloid discs were identical. In order to adjust the pitch to the twelve semitones, transmission pulleys of different sizes and leather belts were used to increase the discs' revolution speed.

A further patent,[14] which Spielmann filed after his immigration to the United States, describes additional sound filters. The apparatus's ability—as announced on a radio show—to reproduce "Fritz Kreisler's violin or Enrico Caruso's voice," including different pitches (Kreisler playing the double-bass or Caruso singing bass),[15] was never accomplished: The instrument held in the collection of the Vienna Museum of Technology comes with two sets of hand-painted sound discs showing sinuslike curves and triangle waves. A reconstruction of the discs for the exhibition *Zauberhafte Klangmaschinen* in the Austrian city of Hainburg (2009) showed, however, that irregularities in the hand-painted curves produce a rather wobbly sound, especially in higher pitches, thus making it rather unsuitable for performing music (Figure 9.2).

In two letters, of June 9 and December 9, 1928,[16] Spielmann invited Erich Wolfgang Korngold (a composer living in Vienna) to take a look at the Superpiano, after Korngold had "considered" presenting the instrument to the public at the Oesterreichische Kulturbund. Enclosed with one of his letters, Spielmann sent an article by music critic Robert Konta[17] describing a demonstration of the instrument to a small circle of professional musicians in Vienna, in which the critic enthuses about the instrument having allowed him to experience Bach, Wagner, and Scarlatti in a new way he had never experienced before. He predicted a "triumphant success" for the Superpiano "around the globe."[18] Korngold indeed presented the instrument in a public concert held on January 9, 1929, at the festival hall of the Kulturbund in Palais Erzherzog Wilhelm in Vienna's city center (Figure 9.3). The music presented included a piece by Korngold himself,

FIGURE 9.2
The Superpiano's sound discs at the Vienna Museum of Technology. Courtesy of Peter Donhauser.

FIGURE 9.3
Korngold at the Superpiano during the presentation of the instrument at the Kulturbund. Courtesy of the Vienna Museum of Technology.

Liebesbriefchen.[19] Contemporary press reports praised the instrument's "inconceivable" possibilities.[20]

A day after the performance, music critic Josef Reitler (an assistant of Korngold's father, Julius, at *Neue Freie Presse*, who was usually in charge of reviewing the performances of his son) wrote that while the instrument might well open up new possibilities that a piano could never offer, it was, however, unfinished, and claims so far made clearly belonged to the realm of fantasy.[21] A month later, a sound film of the Superpiano was shot in Spielmann's apartment on February 7, 1929.[22]

An article in *Neue Freie Presse* infers a link between the Superpiano and Thirring's selenium experiments. Following a brief historical overview, the report refers to Spielmann's "selenium piano," but the main focus is on the transmission of voice by light, starting with Alexander Graham Bell's experiments of 1880. The article describes one of Thirring's demonstration sessions, which included the wireless transmission of a piece of music by means of a ray of light.[23] Earlier, on February 14, 1929, Spielmann had presented his instrument (Figure 9.4) on a Vienna radio (RAVAG) program as part of a series of lectures on "when light speaks, when light makes music." While the first lecture had Hans Thirring talking about "optical telephony," including demonstrations, Spielmann used the second to hail

FIGURE 9.4
The Superpiano at the Vienna Museum of Technology. Courtesy of the Vienna Museum of Technology.

his invention as combining the advantages of the organ, the harmonium, and the piano.[24] *Radiowelt* magazine went even further in its expectations: "That way, we could have Battistini perform a piece of music composed after his death by photographing a single tone from a Battistini record for the instrument and recording it onto the discs."[25]

Subsequently, little more was heard about the Superpiano. The last reference to be found announces a radio program for two Superpianos scheduled for April 8, 1933,[26] presumably including Georg Jokl's[27] "Elegie für 2 Super-Pianos."[28] In 1939, Spielmann left Austria for London, to board the *Veendam* in Southampton on May 6. On August 22, 1944, he applied for U.S. citizenship in New York.[29] His date of death is unknown. The instrument itself was sold to the Vienna Museum of Technology by piano maker Julius Hofmann in 1947.

Further references, whose origin is based on a misreading, mention Thirring in connection with an optical-sound instrument. Meyer-Eppler describes Spielmann's Superpiano[30] but goes on to say: "Thiring's [*sic*] piano is of very similar design," referring to two articles in specialist journals.[31] The author of these articles, Oskar Vierling (who in Berlin in the 1930s was intensively concerned with electronic instruments) does mention the Superpiano, but the article is much more clear-cut about Thirring.[32] He compares

Thirring's construction to one of Emile Hugoniot's experimental instruments:[33] "Thiring [*sic*], on the other hand, built a fully playable version of his Superpiano. Like Cahill, he used 12 discs, each with 7 circles of holes . . . Each row of holes had 7 photo cells and 7 small lamps attached to it." The comparison with Cahill is inappropriate, as Cahill never built an optical-sound instrument. The reference he quotes[34] is based on an article in the Austrian magazine *Radiowelt* reporting on a talk given by Thirring about optical sound transmission. Since Thirring was experimenting with selenium cells at the time[35] and was much better known among experts than Spielmann, the instrument was initially attributed to the university professor.

A design directly comparable to the Superpiano is the Variophone, built by the Russian Evgeny Sholpo in 1931.[36] Sound samples on cardboard discs rotated in a ray of light in order to produce audio-frequency vibrations in a photocell (Figure 9.5). Like many creations by Russian inventors, this instrument is virtually unknown in the Western world. Andrey Smirnov of the Theremin Centre in Moscow offers an explanation: "As a result, in Soviet Russia by the late 1930s, work in this domain effectively came to a halt.

FIGURE 9.5
Sholpo inserts an optical disc into his Variophone. Courtesy of Andrei Smirnov.

The most important documents related to Graphical Sound were never published at all and were circulating in manuscript form, similar to 'Samizdat' (self-published forbidden literature) from the 1960s through the 1980s. Almost no information about existing concepts and inventions was translated and published in any foreign language."[37]

Magneton

In the aftermath of the Great Depression, and as a consequence of changing leisure activities thanks to the diffusion process of new entertainment media such as film and radio in the late 1920s, the musical instrument industry, and especially the piano manufacturing industry, faced serious financial trouble. It is thus little surprising that the sector was looking for new products and markets. Examples include the Neo-Bechstein piano, the Förster Elektrochord, and the Welte Lichttonorgel. In Austria, the renowned piano maker Rudolf Stelzhammer decided to try his luck with electronic organs. As the company lacked the necessary know-how, Stelzhammer acquired a patent held by Wilhelm Lenk, an engineer from Linz and assistant at the Phonetics Department of Vienna University.[38] Lenk's design featured toothed *anchor discs* rotating at different speeds (corresponding to equal temperament tuning) in front of coils in order to generate an audio-frequency alternating current by means of changes in the magnetic flow.[39] This ultimately goes back to the invention of Thaddeus Cahill.[40] As so often, Lenk's idea was not the only one of its kind: six months before him Oskar Vierling patented a similar design,[41] and further patents can be found in Europe and the United States. Laurens Hammond's design was patented four years after Lenk's.

It is a long way, however, from the initial idea to a playable instrument. In 1930, Stelzhammer commissioned a man named Max Pichl to carry out the design and construction of the first Magneton. A special department was formed by Stelzhammer for the project,[42] and a press event was arranged to demonstrate an experimental model.[43] Three years later, however, Stelzhammer and Pichl had come to blows and ultimately broke off their collaboration: Stelzhammer accused Pichl of faulty designs, of announcing excessive prices for the instrument and, ultimately, of not accomplishing the construction of a saleable instrument.

At that stage, Ernst Werndl appeared on the scene. The nephew of the eminent Steyr/Upper Austria industrialist Josef Werndl, he had failed to make a career for himself in his uncle's company or anywhere else. After three years of studying electrical and mechanical engineering in Vienna, Ernst Werndl left for America, where he briefly worked in Edison's laboratory. Back in Vienna, he went back to the university to read physics for some time, before taking off again and moving from one city to another almost annually for some years. In 1932 he finally returned to Vienna, where he lived in dire straits. He even had to put his furniture in storage with Stelzhammer for a while and live on the factory grounds. Pressured by his wife, on July 15, 1933, he sold his patent for a "contact apparatus" to Stelzhammer, who had already acquired Lenk's patent. In 1935 Werndl managed to stop Pichl from taking out a patent for a driving unit for the Magneton, claiming the sound absorption system it described to be his own idea (Figure 9.6).

Werndl's diaries provide detailed information on the instrument's further development. The first reference to the Magneton can be found in an entry dating back to November 19, 1932: a vibrato by means of an excentric within the driving unit of the tooth wheels and an optical indicator for adjusting the timbre. On November 20, 1932, he finalized his concept for a contact apparatus for the Magneton, and he filed a patent five days later.[44] He was pondering a "pure-tone function" and, in connection with it, a simple tuning device as well as the possibility of third-tones. In February 1933 Werndl was still pleased with the Magneton's Principal (name of a pipe organ stop) sound, but he writes on November 4, 1933: "Magneton works; am still not content with it"; and on November 6, "mixtures essential" (in any case, these are not part of the extant instrument). Numerous additional ideas followed, including, for instance, separate amplifiers for each partial. One entry in particular might be useful for reconstructing the instrument's register function: "multiplying" register switches (each increasing the amplification factor's exponent by one).[45] From time to time, Werndl returns to an older idea, the Optophon (with twelve endless tapes of sound

FIGURE 9.6
The Magneton's sound discs. Courtesy of Peter Donhauser.

patterns),[46] but he keeps coming back to the Magneton and, as an alternative, suggests an instrument fitted with twelve records whose endless tracks provide the required octaves and register timbres. In Werndl's view, the Stelzhammer Magneton might be perfect as a church organ complementing the major pipe organ, as a stage or radio organ, or as a practice organ for use at home (Figure 9.7).

When the construction of the instrument had dragged on to its completion, Stelzhammer began to look for marketing opportunities. Plans for a new church building in Vienna, the "Dr. Ignaz Seipel Memorial Church," designed by the prominent Austrian architect Clemens Holzmeister (one of his famous buildings is the Great Hall of the Salzburg Festival) in 1933, seemed a promising opportunity.[47]

Werndl's proposals for the organ were ambitious: two consoles (one in the gallery, one within the church's main hall) were to be used to play both a pipe organ and the Magneton. The project was prestigious as it was initiated by Hildegard Burjan, the cofounder, with Ignaz Seipel, of the charitable religious congregation of Caritas Socialis and a Christian-Social Party member of the Austrian parliament in Vienna. After a

FIGURE 9.7
The completed Magneton. Courtesy of the Vienna Museum of Technology.

disappointing presentation of the Magneton on the radio, however, which Stelzhammer once again blamed on Pichl, the contract fell through.[48] Obtaining additional expert recommendations was impossible: the head organist of St. Stephen's Cathedral in Vienna was bound to another organ manufacturer, Marcel Kaufmann; Prof. Karl Walter, head of the Department of Church Music at the Vienna Academy of Music, was generally opposed to any music based on the electrical generation of sound. In 1937, the church finally acquired a Kaufmann pipe organ.[49]

The last public performance of the Magneton in Austria took place at the Vienna Urania on December 4, 1934, in an evening event titled "New liturgical music performed on a modern instrument." This "promising beginning"[50] was followed by another short-lived period of success, but eventually work on the instrument was abandoned.

In 1935 Stelzhammer took part in the eleventh International Exhibition of Inventions[51] in London, where the Magneton outdid five hundred competitors to win first prize.[52] For his achievement, Stelzhammer was (as his nephew, Hugo Stelzhammer, recalled in a conversation with me) awarded the title of professor. According to a news comment,[53] architect Clemens Holzmeister had praised the instrument as a way to resolve acoustic problems in church buildings. After this London success it was claimed that the Magneton was suitable even for St. Paul's Cathedral, and that out in the open, it was effective over a distance of 1,500 meters. It could be successfully played in Hyde Park.

Werndl himself continued taking notes of ideas for the Magneton until 1935, including an improved driving unit[54] and modified sound discs. By that time he had apparently left Stelzhammer's, as the same year he had a certain Manfred Zeilinger to file three patents for the improved sound discs in Salzburg.[55] None of these designs, however, were ever realized. Werndl only heard about the Magneton's London success through one of his uncle's letters.

In 1973, the Vienna Museum of Technology acquired the instrument from Klavierhaus Hugo Stelzhammer. A few months later, however, it was so fundamentally altered during attempts to repair it (among other components, the contact apparatus was removed) that a successful reconstruction of the instrument is unlikely (Figure 9.8).

Variacord

Werndl's contributions to instrument building in Vienna also include the Variacord, designed by electrical engineer Robert Pollak-Rudin,[56] who worked in electroacoustics for some time, in particular in the production of sound records[57] (he briefly had his own record label, Tilophan). He ran several recording studios in Vienna, Innsbruck, Graz, Linz, and Salzburg and produced recordings for amateur musicians. He organized public performances to market his technology.[58] He also produced "minus one" records.[59] Pollak-Rudin was also interested in electronic musical instruments and received several patents for his designs. Werndl first mentions being in contact with him on February 21, 1935,[60] and they filed their first joint patent in 1936[61] for an instrument based on optical sound.

FIGURE 9.8
Werndl's contact apparatus for the Magneton. Courtesy of Peter Donhauser.

The mechanism differed from Pollak-Rudin's other designs insofar as it featured a rotating diaphragm and fixed sound waves rather than the other way around, as was usual at the time. No prototype seems to have been built, as Werndl's (very detailed) diaries make no reference to it.

That was not the case with the following patent:[62] an electromagnetic string instrument called the Variacord. An electric impulse from a capacitor discharge, run through a magnet, was to briefly excite the string, in a similar way to a piano hammer. Different magnet positions were to allow for different timbres. To vary the dynamics, the capacitor's charging voltage was adjusted (Figure 9.9). Incorporating the contact apparatus Werndl had developed for the Magneton allowed for various coupling possibilities, such octave and suboctave couplers.[63]

A prototype of the instrument described in the patent was actually built and presented at the Grosse Ehrbar-Saal in Vienna,[64] which Pollak-Rudin had rented for the occasion[65] on October 26, 1937. Two newspapers[66] mention (but do not judge) the performance (Figure 9.10); and, according to Werndl's diaries, it was, at any rate, not a failure.[67] Pollak-Rudin's son thinks he remembers excellent reviews.[68]

FIGURE 9.9
The open Variacord. The amplifier (top left); the contact apparatus (top right); the strings (underneath). Source: *Radio Amateur*, July 14, 1937.

Although the pianist Rudolf Serkin would have nothing to do with the instrument, other young musicians, including Renée Gärtner, orchestra conductor Heinrich Krips,[69] H. Vogel, and Jakob Gimpel played it in public.[70] The instrument was said to be perfectly suited to solo performances as well as to jazz music and symphonic orchestras. Interestingly, some Tilophan recordings were actually dedicated to jazz,[71] which Pollak-Rudin apparently had an interest in.

When the Variacord's inventor was forced to leave Austria in 1939, and Werndl went first to Telefunken in Berlin in 1938 and then to join Vierling in Ebermannstadt[72] in 1943,[73] all went quiet about the instrument. After the end of the war, Werndl tried to resume his experiments with electronic instruments with a company in Salzburg, but the workshop burned down. Werndl's attempts to obtain a professorship for electronic music in Salzburg also failed. The Variacord was last heard of in a radio lecture given

FIGURE 9.10
Werndl playing the Variacord; Pollak-Rudin standing behind him. Source: *Kleines Volksblatt*, November 2, 1937.

by Werndl, broadcast by Rot-Weiss-Rot radio on December 4, 1946. The program also included samples from a record of Variacord music.[74] Unfortunately, both the record and the instrument are missing. The instrument is nevertheless relatively well known, as it was described in a dissertation submitted to Humboldt-Universität Berlin in 1952[75] and is also referred to in Ferdinand Scheminzky's *Die Welt des Schalls* from 1943.[76]

Chromatophon

The synaesthetic[77] experiments to translate music into visual imagery popular during the 1920s and 1930s also include a little-known venture undertaken by the Baltic pianist Baron Anatol Vietinghoff-Scheel[78] in collaboration with Josef Kanzler, a blind piano maker from Graz, and several electrical engineers. The entire project, from construction to public demonstration, was financed by Kanzler, almost ruining him.[79] Kanzler had an Ehrbar concert grand piano adapted and fitted with a series of switches for colored lights and projectors. The instrument was named the Chromatophon (Figure 9.11).

FIGURE 9.11
The adapted Ehrbar grand on stage at the Grosse Saal of the Musikverein in Vienna. Reproduced with permission from private owner.

The stage background was fitted with an art deco backdrop of sorts (Künstlerische Raumgestaltung) designed by painter Franz Koeck and incorporating a multitude of colored lights controlled by the switches on the Chromatophon piano. The backdrop ("a weightless composition of cylinders, prisms and cones")[80] measured as much as ten meters in width, seven meters in height, and four meters in depth. Anatol Vietinghoff-Scheel had devised a system in which each tone was associated with a clearly defined color:

C	bluish white
C sharp = D flat	dark purple, with a tendency toward blue for C sharp and red for D flat
D	indigo
E flat	light blue
E	green
F	red
F sharp = G flat	mix of red, blue, and brown, with brown as the dominant color and a tendency toward blue for F sharp and red for G flat
G	black, brightened to grey in lit conditions

A flat = G sharp	lilac, with G sharp producing a bluish tint
A	blue
B flat	orange
B	olive

A complex system controlled the lights; each of the Chromatophon's keys was connected to one of seventy-two small spotlights, their brightness adjusted by a foot-controlled resistor underneath the instrument. The sustain pedal kept the lights aglow after the key was released. Toward the front of the stage, another set of lights, covered by a plate of milk glass, faced the audience. A third set (ground-level lighting) was available to visualize the most important keys, controlled by twenty-four pedals underneath the piano and dimmed with the help of a roller (Figure 9.12). Each "piece of sound art" was assigned to its own décor, some featuring bizarre shapes or veils. An in-built film projector was used to project ephemeral shapes and colors onto the scenario, illustrating impressionist pieces of piano music. A third roller controlled its luminosity. Altogether, it had a respectable power consumption of 25 kW.[81]

The apparatus had its public debut on December 7, 1929, at the Industriehalle in Graz (Styria). It was a great success,[82] despite Kanzler's considerable financial losses. In August that year, the first "visual music" experiments had been conducted at the Graz

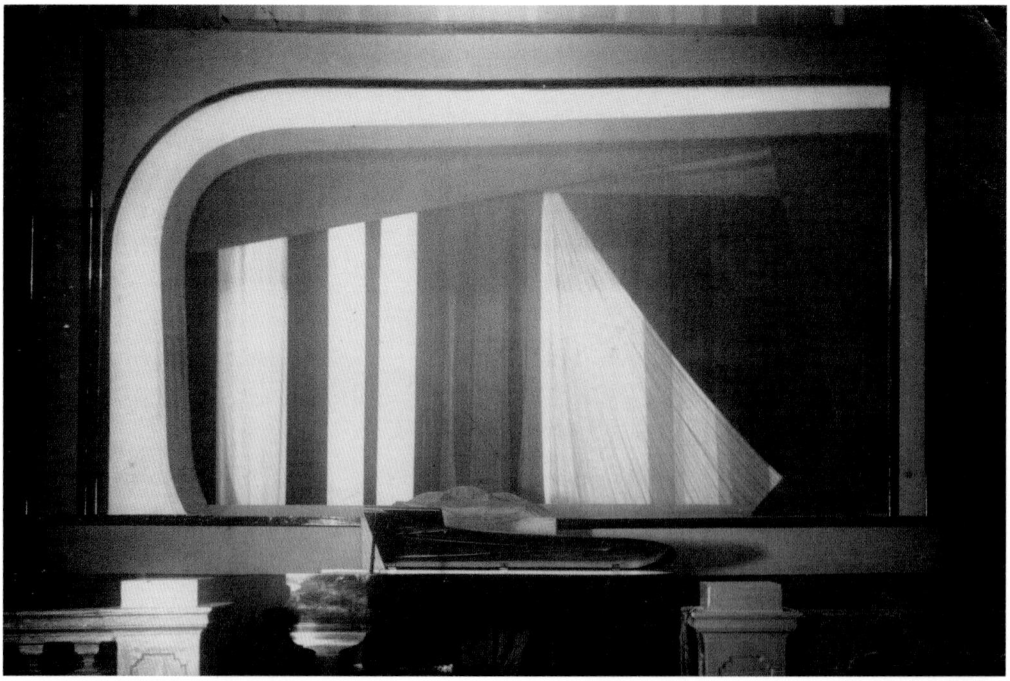

FIGURE 9.12
The Chromatophon stage. Reproduced with permission from private owner.

Opera House. A public performance was originally planned for November 29, 1929, at the Grosse Saal of Vienna's Konzerthaus, but it was cancelled the week before and moved to the Grosse Saal of Musikverein in Vienna on December 12, 1929. In contrast to the Graz performance, the reviews it got here were lukewarm to disastrous. Robert Konta (apparently blissfully unaware of the phenomenon of synaesthesia) scathingly commented that any boxer receiving a blow to the head was likely to see colors as well. The effect the music might have on overly sensitive (perhaps even pathologically nervous) people may well be similar to a punch in boxing, "literally leaving a lasting impression." His rejection of performances like this, he claimed, was not as much backward as farsighted.[83] *Kleine Volkszeitung* put it even more bluntly: "As these days anyone seems to be free to do whatever they want on German stages, we too now find stages draped in sheets of all sorts, of which those in dark hues—the colour of mourning—make the most sense." The article went on to maintain that it was impossible to make out a connection between the music and colors; Vietinghoff-Scheel would be better off sticking to playing the piano (Figure 9.13). Rather than musical brilliance, this was little more than technical shenanigans. "After the first couple of pieces, all had vanished. Not before a trip to the cloakroom, that is."[84] Other critics were more dispassionate but at best noncommittal.[85]

At the second Colour-Music Congress hosted by Hamburg University in 1930, Vietinghoff-Scheel gave a lecture on the Chromatophon[86] but lacked the funds for a

FIGURE 9.13
Anatol Vietinghoff-Scheel in a newspaper clipping. Source: *Neues Wiener Abendblatt*, December 21, 1929.

demonstration. He died three years later, without further promotion of his arrangement. The altered Ehrbar grand is missing, considered lost.

Incidentally, the Czech sculptor, Zdeněk Pešánek, was experimenting with the visual translation of music into colors at the time (Figure 9.14). He also presented a paper at the Hamburg Congress.[87] The *Berliner Tageblatt und Handelszeitung* newspaper reported on it in its issue of June 15, 1930, also with some skepticism: "This may well lead to little more than new gimmicks for illuminated advertising. But it is also possible that it opens up entirely new paths of artistic expression."[88]

Heliophon

The Heliophon's origins go back to a design devised in Frankfurt in the late 1920s. It is mentioned here for the sake of completeness, as it was in use in Austria after the end of the Second World War.

FIGURE 9.14
Pešánek's apparatus for color music, set up at the Edison transformer station in Prague. Source: Georg Anschütz, *Farbe-Ton-Forschung* (volume 3, Hamburg, 1931).

Inspired by Lev Termen's Aetherphone, which was first presented in Germany at that time, a pianist from Frankfurt, Bruno Helberger, and physicist Peter Lertes, collaborated in the development of their own electronic instrument called the Hellertion, which featured a series of fingerboard strips.[89] The design of the fingerboards, which allowed for continuous glides between pitches, not only resembles the manual of Trautwein's Trautonium but also anticipated the ribbon controllers of electronic musical instruments of later decades. Although no effort was spared, and the partners even had contracts with Telefunken to start commercial production, the project fell through.[90]

To help establish his instrument, Helberger went as far as chumming up to the Nazi regime and contributing to National Socialist Party events. Apparently, though, toward the end of the war, Helberger decided to leave Germany and went to stay with relatives in Carinthia, Austria, where he had also been registered as a resident since 1939. During this period he modified the Hellertion, incorporating a keyboard and renaming the instrument Heliophon. He built two different-sized prototypes. In September 1947, composer Joseph Marx went to Carinthia to view the instrument. He was especially taken with the possibility of playing two timbres on a single manual.[91] This was achieved by means of a variable split point (Figure 9.15).

Helberger took part in several public engagements, including performances at the Volkstheater and the Burgtheater in Vienna, the Vienna State Opera, as well as several radio shows after 1945 and in the 1950s. He collaborated with composer Paul Angerer and contributed to early TV films. His last instrument was acquired by his brother-in-law before Helberger's death in 1961. On the initiative of a student of Helberger's son, Heinz-Peter, the instrument was moved from Klosterneuburg to Berlin for the exhibition *Auf der Suche nach dem neuen Klang* in 1980. It was subsequently donated to the Berlin Museum of Musical Instruments, where it is currently in storage.

Outlook

What all these instruments had in common was that they required a musician or "interpreter"; that is, someone to play or "operate" them. The new electronic sound was frequently used to interpret classical music or to provide conventional musical entertainment (the latter continuing to this day). Only in France and Russia did electronic instruments inspire composers to create new pieces; Paul Hindemith's interest ultimately waned after he left Nazi Germany (whose idea of culture would have disallowed experimental music anyway).

The coming to power of National Socialism affected the development of electronic music in Germany and Austria in different ways. In Germany there was some political interest in the development of electronic musical instruments, even though the National Socialists opposed contemporary music. In Austria the new political situation had the effect that two Jewish people deeply involved in the technology (Spielmann and Pollak-Rudin) emigrated (Stelzhammer's experiments had failed before Austria's occupation by

FIGURE 9.15
The smaller Heliophon model, which Joseph Marx viewed. Source: *Österreichische Musikzeitschrift* 1947, Heft 11/12.

the Nazis in 1938). As in Germany, the Second World War put a definitive end to individual development of such instruments in Austria. The electronic instruments in use after the end of the war were imports, mainly from the United States, Great Britain, and later on from Germany.

After the end of the war, an entirely new approach to electronic music emerged. Not just memories of the futurists' "noise music" of the 1920s, but also the new possibilities of electronic sound production (hitherto not sufficiently used) gave rise to an entirely new experimental music scene. The ground for such developments was prepared by new electronic studios, such as those in Cologne, Munich, and East Berlin. These produced entirely novel forms of music: generators, tape recorders, and records now were the sources of sound. Music no longer required a musician in the conventional sense of the word.

No similar institutions were available in Austria, so the development of electronic music proceeded very slowly. Here, it was at music universities that experimental instruments were developed over the next thirty years, such as the Akaphon (1963), the AKA 2000 (1975–1985) in Vienna, and the Hoenig Synthesizer (since 1965) in Graz. These instruments are now on display at the Vienna Museum of Technology in a permanent exhibition devoted to synthesizers.

Globalized music production, the ready availability of a vast range of synthesizers and the rise of computer and music software in electronic music production notwithstanding, we should not forget that the roots of these developments go back more than a hundred years and are based on simple principles of electrical engineering.

Notes

1. The detailed history of these instruments is documented in existing literature; for example, Peter Donhauser, *Elektrische Klangmaschinen. Die Pionierzeit in Deutschland und Österreich* (Vienna: Böhlau, 2007).

2. Frode Weium, "Ingenieurmusik. Die Begegnung mit elektrischen Musikinstrumenten in Norwegen in den 20er- und 30er-Jahren," *Blätter für Technikgeschichte* 69/70 (2008): 135ff.

3. "Seraphicum," recorded on September 12, 1935, by the German Reichsrundfunkgesellschaft, original lost. See Donhauser, *Elektrische Klangmaschinen*, appendix. Copy on tape in the estate of Werner Meyer-Eppler, Akademie der Künste, Berlin. Thanks to Caspar Oehlschlägel for the hint.

4. Fred K. Prieberg, *Musik im NS-Staat* (Frankfurt am Main: Fischer Taschenbuch Verlag, 1982). Christine Fischer-Defoy, *Kunst, Macht, Politik: Die Nazifizierung der Kunst-und Musikhochschulen in Berlin* (Berlin: Elefanten Press, 1988). Albert Glinsky, *Theremin. Ether Music and Espionage* (Chicago: University of Illinois Press, 2005).

5. Anonymous, "Von Forschergeist und Alt=Wiener Werkmannsarbeit," *Neue Freie Presse*, August 14, 1930.

6. Anonymous, "Ganz neue wissenschaftliche Wege werden da zum ersten Mal betreten," *Neues Wiener Extrablatt*, December 10, 1929.

7. Donhauser, *Elektrische Klangmaschinen*.

8. Photographischer Registrirapparat für telephonische Übertragung, patent DE 27.231, dated September 4, 1883.

9. Patent GB 10.363: Improvements in and Relating to Multiple Telegraphy. The different tones are used for signal transmission in telegraphy. Mercadier filed another two improvements: GB 13.322 (1889) and GB 21.357 (1895).

10. Anonymous, *The Superpiano*, Brochure in English without specification. Vienna Museum of Technology archive.

11. Patent DE 443.535.

12. *Zeitschrift für Instrumentenbau*, Volume 1926/27, 447.

13. Elektrisches Musikinstrument, Patent AT 109.233 filed on February 11, 1927.

14. Patent US 2,469.850.

15. Text of the broadcast on February 14, 1929, RAVAG. See Anonymous, "Das Geheimnis der Elektromusik," *Neues Wiener Journal*, January 9, 1929, 5.

16. Austrian National Library, manuscript collection, shelfmarks 941/67-1.Han and 67-2. Han. Both letters hail from an omnibus volume, which the Austrian gestapo confiscated in Korngold's domicile in 1938 and transferred to the Austrian National Library.

17. Austrian composer, music writer, and music critic, 1880–1953.

18. Attachment to 941/67-1 without date and source.

19. Anonymous, "Theater-und Kunstnachrichten. Das Superpiano," *Neue Freie Presse*, January 10, 1929, 9.

20. Anonymous, "Das Superpiano," *Neue Freie Presse*, January 9, 1929, 3.

21. Anonymous, "Theater-und Kunstnachrichten."

22. The Vienna-based Movietone group of journalists filmed the instrument. The whereabouts of the footage today, however, are unknown. Anonymous, "Mitteilungen aus der Filmindustrie," *Neue Freie Presse*, February 8, 1929, 7.

23. Anonymous, "Tönende Zelle und sprechender Strahl," *Neue Freie Presse*, February 22, 1929, 8.

24. "Das Licht spricht, das Licht musiziert," RAVAG, broadcast on February 14, 1929.

25. Anonymous, "Das Spielmannsche Lichtklavier," *Radiowelt* 3 (1929): 73.

26. Anonymous, *Rundfunk*, 1933: 14th week, 14.

27. Viennese composer, 1896–1954. A student of Franz Schreker, he was prosecuted by the National Socialists and immigrated to New York.

28. Musical collection of the National Library, shelfmark Mus.Hs.40642. a,b Mus.

29. Passenger list of the vessel and the application form.

30. Werner Meyer-Eppler, *Elektrische Klangerzeugung* (Bonn: Duemmler, 1949), 112.

31. Oskar Vierling, "Elektrische Musik," *ETZ* 53 (1932): 155ff, and Oskar Vierling, "Das elektrische Musikinstrument," *ZsVDI* 76(32) (1932): 743.

32. Ibid.

33. Patent Instrument de musique électrique, FR 550.370.

34. "Revue francaise de T.S.F.," February 1929, 12 (an amateur radio magazine). The note was also published in Robert de Valbreuze, *Comptes rendus du Congrès International d'Électricité* (Paris: Gauthier-Villars, 1932), 195.

35. Patent AT 104.324.

36. "Вариофон Евгения Шолпо." http://www.theremin.ru/archive/variophone.htm (accessed October 24, 2009).

37. See http://asmir.theremin.ru/gsound_concert1.htm (accessed October 24, 2009).

38. Anonymous, "Magneton. Eine neue Wiener Erfindung," *Neue Freie Presse*, August 14, 1930, 8.

39. Patent AT 128.615, dated March 29, 1930.

40. Cahill (1867–1934) got his first patent (US 580.035) in 1897. The specification comprised a carefully elaborated system, including many drawings and circuit diagrams for an instrument with rotating tone generators. A detailed description has been published by Reynold H. Weidenaar, *The Telharmonium: A History of the First Music Synthesizer, 1893–1918*, Dissertation, New York University, 1989.

41. Patent DE 529.948, dated November 22, 1929.

42. "Übereinkommen," estate Stelzhammer.

43. Anonymous, "Magneton. Eine neue Wiener Erfindung," *Neue Freie Presse.*

44. Patent AT 135.789.

45. All of this information comes from Ernst Werndl, *Diaries*, Upper Austrian province archive.

46. Werndl, *Diaries*, Volume XIX, January 6, 1934, 155. Werndl's concept closely matches that of patent US 2,030.248 filed by Ivan Eremeeff, whose idea for synthesized sound, even including subharmonics(!), goes far beyond Werndl's and even anticipates Oskar Sala's idea for his Mixtur-Trautonium. It goes back to at least 1932, when he tried to interest radio pioneers Oskar Czeija and Franz Nissl in his idea.

47. Today known as Christkönigskirche, 1150 Vienna, Vogelweidplatz 7–8, consecrated September 29, 1934.

48. Letter to Max Pichl, dated November 10, 1934, estate Stelzhammer, private ownership.

49. Oskar Eberstaller, *Orgeln und Orgelbau in Österreich* (Graz: Böhlau, 1955).

50. Anonymous, "Das Magneton," *Musikblätter des Anbruch* 16 (1934): 202.

51. F. W. Galpin, "The Music of Electricity," *Proceedings of the Musical Association*, February 17, 1938, 78.

52. Anonymous, "Wiener Erfinder in London preisgekrönt," *Neues Wiener Journal*, October 20, 1935, 20. The article denotes the event as "Erfinder-Konkurrenz."

53. Anonymous, "Wiener Erfinder in London preisgekrönt," 20.

54. Werndl, *Diaries*, Volume XXII, entry of September 26, 1935.

55. Patents AT 154.277, AT 152.069, and AT 151.153. Zeilinger himself is not notable except for three further patents about dentistry.

56. Born in Vienna in 1891. In 1914, he gained a doctorate from Vienna University of Technology; 1927, court expert for sound film and radio; 1939, immigrated to Switzerland and, later on, to France (meeting with Givelet); 1941, immigrated to the United States: died in New York in 1957.

57. Patent AT 149.202, Verfahren zur Herstellung eines zum unmittelbaren Einschneiden von Schallnuten geeigneten Schallschriftträgers.

58. For example, on November 7, 1935. See Anonymous, "Moderne Schallplattenaufnahme," *Die Stimme*, November 5, 1935, 4.

59. V. Rudin-Usdin, *I Remember It Differently*. Unpublished memoirs.

60. Werndl, *Diaries*, Volume XXII, 254.

61. Patent AT 153.659, Verfahren zur Erzeugung von Tönen oder Klängen auf lichtelektrischem Wege.

62. Patent AT 153.660, Verfahren zur Tonerzeugung, insbesondere bei elektrischen Musikinstrumenten.

63. Anonymous, "Variacord. Ein neues elektrisches Musikinstrument," *Radio-Amateur*, July 14, 1937, 407.

64. Erected in Mühlgasse 30, in Vienna's 4th district, 1867, by piano manufacturer Friedrich Ehrbar. Between 1934 and 1938, the Konzerte Musik der Gegenwart series of concerts related to Marcel Rubin was held there.

65. Walter Rudin, *The Way I Remember It* (American Mathematical Society, 1997), 22.

66. Anonymous, "Ein neues Musikinstrument," *Kleines Volksblatt*, November 2, 1937, and Anonymous, "Das neue Klavier Variacord," *Telegraf am Mittag*, October 27, 1937. Both are attachments to Werndl's diaries.

67. Werndl, *Diaries*, Volume XXIV, entry of October 26, 1937.

68. See Rudin, *The Way I Remember It*.

69. Both names are mentioned on an invitation card, estate Werndl.

70. Rudin-Usdin, *I Remember It Differently*, 43.

71. Anonymous, *Discographie der Österreichischen Populärmusik*, Phonomuseum Vienna, without date.

72. In 1941, Oskar Vierling had a building erected on his own land, financed by the Wehrmacht and disguised as a "castle," where he developed underwater microphones and high-frequency technology for the Wehrmacht. For details, see Donhauser, *Elektrische Klangmaschinen*, 233.

73. Vita, summer 1947, typescript in estate Werndl.

74. Letter by Wendl to Dr. Burstyn, dated February 8, 1947, estate Werndl.

75. Erich Stockmann, *Der musikalische Sinn der elektro-akustischen Musikinstrumente* (Dissertation, Berlin University, 1952), 29.

76. Ferdinand Scheminzky, *Die Welt des Schalls* (Salzburg: Verlag Das Berglandbuch, 1943), 738.

77. Synaesthesia is a neurologically based condition in which stimulation of one sense leads to automatic, involuntary experiences in a second sense. Sound-to-color synaesthesia occurs when voice, music, and assorted environmental stimuli trigger colors and simple shapes that arise, move around, and then fade when the sound stimulus ends.

78. Born in 1899 (in Tsarskoye Selo near St. Petersburg), Baron Anatol Vietinghoff-Scheel died in 1933 (in Graz). He studied at St. Petersburg Conservatory and made concert tours in Japan and Western Europe. In 1924, he settled in Graz as a music teacher. His works include a musical drama, two operas, two symphonies, three piano concertos, and others (see Alexandria Vodarsky-Shiraeff, *Russian Composers and Musicians* [New York: Kosta Press, 1940], 142). According to Vietinghoff-Scheel himself, the "concert tours" he undertook were really an attempt to flee from the Bolsheviks and, later, the necessity to leave Japan after a major earthquake. Following a visit to a health spa in Bad Gleichenberg, Styria, he decided to stay in Styria. (From Anonymous, "Farbenmusik," *Wiener Tagblatt*, November 16, 1929.)

79. The material is the private property of Kanzler's descendants.

80. Anonymous, "Aus dem Konzertsaal," *Grazer Montagszeitung*, December 9, 1929.

81. Georg Anschütz, *Farbe-Ton-Forschung* Vol. 3 (Hamburg: Psychologisch=aesthetische Forschungsgesellschaft, 1931), 389.

82. Anonymous, "Musik in Farben," *Grazer Tagespost*, December 9, 1929, and Anonymous, "Das Chromatophon," *Grazer Tagblatt*, December 10, 1929.

83. Anonymous, "Farben hören, Töne sehen," *Wiener Allgemeine Zeitung*, December 18, 1929.

84. Anonymous, "Musik für nichts," *Kleine Volkszeitung*, December 17, 1929.

85. For example, Anonymous, "Sichtbare Musik," *Neue Freie Presse*, December 14, 1929.

86. Anschütz, *Farbe-Ton-Forschung.*

87. Ibid., 193.

88. Anonymous, "Lichtmusik? Ein neuer Versuch, auf Grund der Elektrotechnik eine neue Kunstform zu schaffen," *Berliner Tageblatt und Handelszeitung*, June 15, 1930, 6.

89. Musikinstrument, bei dem Tonfolgen durch Frequenzänderungen eines Röhrenoszillators erzeugt warden, patent DE 549.481, dated December 20, 1928, with addendum DE 552.040, dated May 27, 1929.

90. Numerous details in Donhauser, *Elektrische Klangmaschinen.*

91. J. Marx, "Heliophon, ein neues Musikinstrument," *Österreichische Musikzeitschrift* Vol. 11/12 (1947): 314.

Bibliography

Anonymous. "Aus dem Konzertsaal." *Grazer Montagszeitung*, December 9, 1929.

———. "Das Chromatophon." *Grazer Tagblatt*, December 10, 1929.

———. "Das Geheimnis der Elektromusik." *Neues Wiener Journal*, January 9, 1929, 5.
———. "Das Magneton." *Musikblätter des Anbruch* 16 (1934): 202.
———. "Das neue Klavier Variacord." *Telegraf am Mittag*, October 27, 1937.
———. "Das Spielmannsche Lichtklavier." *Radiowelt* 3 (1929): 73.
———. "Das Superpiano." *Neue Freie Presse*, January 9, 1929, 3.
———. *Discographie der Österreichischen Populärmusik*. Phonomuseum Vienna, without date.
———. "Ein neues Musikinstrument." *Kleines Volksblatt*, November 2, 1937.
———. "Farben hören, Töne sehen." *Wiener Allgemeine Zeitung*, December 18, 1929.
———. "Farbenmusik." *Wiener Tagblatt*, November 16, 1929.
———. "Ganz neue wissenschaftliche Wege werden da zum ersten Mal betreten." *Neues Wiener Extrablatt*, December 10, 1929.
———, "Lichtmusik? Ein neuer Versuch, auf Grund der Elektrotechnik eine neue Kunstform zu schaffen," *Berliner Tageblatt und Handelszeitung*, June 15, 1930, 6.
———. "Magneton. Eine neue Wiener Erfindung." *Neue Freie Presse*, August 14, 1930: 8.
———. "Magneton. Eine neue Wiener Erfindung." *Neues Wiener Tagblatt*, August 27, 1930.
———. "Mitteilungen aus der Filmindustrie." *Neue Freie Presse*, February 8, 1929, 7.
———. "Moderne Schallplattenaufnahme." *Die Stimme*, November 5, 1935, 4.
———. "Musik für nichts." *Kleine Volkszeitung*, December 17, 1929.
———. "Musik in Farben." *Grazer Tagespost*, December 9, 1929.
———. *Rundfunk*. 1933: 14th week.
———. "Sichtbare Musik." *Neue Freie Presse*, December 14, 1929.
———. *The Superpiano*. Brochure in English without specification. Vienna Museum of Technology archive.
———. "Theater-und Kunstnachrichten. Das Superpiano." *Neue Freie Presse*, January 10, 1929, 9.
———. "Tönende Zelle und sprechender Strahl." *Neue Freie Presse*, February 22, 1929, 8.
———. "Variacord. Ein neues elektrisches Musikinstrument." *Radio-Amateur*, July 14, 1937, 407.
———. "Von Forschergeist und Alt=Wiener Werkmannsarbeit." *Neue Freie Presse*, August 14, 1930.
———. "Wiener Erfinder in London preisgekrönt," *Neues Wiener Journal*, October 20, 1935, 20.
Anschütz, Georg. *Farbe-Ton-Forschung* Vol. 3. Hamburg: Psychologisch=aesthetische Forschungsgesellschaft, 1931.
de Valbreuze, Robert. *Comptes rendus du Congrès International d'Électricité*. Paris: Gauthier-Villars, 1932.
Donhauser, Peter. *Elektrische Klangmaschinen. Die Pionierzeit in Deutschland und Österreich.* Vienna: Böhlau, 2007.
Eberstaller, Oskar. *Orgeln und Orgelbau in Österreich.* Graz: Böhlau, 1955.
Fischer-Defoy, Christine. *Kunst, Macht, Politik: Die Nazifizierung der Kunst-und Musikhochschulen in Berlin.* Berlin: Elefanten Press, 1988.
Galpin, F. W. "The Music of Electricity." *Proceedings of the Musical Association*, February 17, 1938, 78.
Glinsky, Albert. *Theremin. Ether Music and Espionage.* Chicago: University of Illinois Press, 2005.
Marx, J. "Heliophon, ein neues Musikinstrument." *Österreichische Musikzeitschrift* Vol. 11/12, 1947, 314.
Meyer-Eppler, Werner. *Elektrische Klangerzeugung.* Bonn: Duemmler, 1949, 112.

Prieberg, Fred K. *Musik im NS-Staat.* Frankfurt am Main: Fischer Taschenbuch Verlag, 1982.
"Revue francaise de T.S.F.," February 1929, 12
Rudin, Walter. *The Way I Remember It.* American Mathematical Society, 1997.
Rudin-Usdin, V. *I Remember It Differently.* Unpublished memoirs.
Scheminzky, Ferdinand. *Die Welt des Schalls.* Salzburg: Verlag Das Berglandbuch, 1943.
Stockmann, Erich. *Der musikalische Sinn der elektro-akustischen Musikinstrumente.* Dissertation, Berlin University, 1952.
Vierling, Oskar. "Das elektrische Musikinstrument." *ZsVDI* 76(32) (1932): 743.
———. "Elektrische Musik." *ETZ* 53 (1932): 155.
Vodarsky-Shiraeff, Alexandria. *Russian Composers and Musicians.* New York: Kosta Press, 1940.
Weidenaar, Reynold H. *The Telharmonium: A History of the First Music Synthesizer, 1893–1918.* Dissertation, New York University, 1989.
Weium, Frode. "Ingenieurmusik. Die Begegnung mit elektrischen Musikinstrumenten in Norwegen in den 20er- und 30er-Jahren." *Blätter für Technikgeschichte* 69/70 (2008): 135–54.
Werndl, Ernst. *Diaries.* Upper Austrian province archive.
Zeitschrift für Instrumentenbau, Volume 1926/27.

CHAPTER 10

New and Rediscovered Musical Instruments Rediscovered

David Toop

IN 1974, immersed in London's emergent improvised music and sound art scenes, I edited a small book titled *New/Rediscovered Musical Instruments*.[1] An anthology documenting experiments in designing and building new audio technology by Hugh Davies, Max Eastley, Paul Burwell, Evan Parker, Paul Lytton, and me, the book's genesis lay in a number of different sources. These reflected both my obsessions of the time and the prevailing discourse developing around the limitations of both conventional instruments and new electronic technology. Much of this discourse has been forgotten, but it may be worthwhile to exhume its central tenets to trace connections between experiments in live electronic music, improvisation, and a parallel growth of interest in ethnomusicology and organology (the analytical and descriptive study of musical instruments) in the 1970s.

In Britain, the majority of 1960s improvisers had roots in modern jazz and more specifically, the free jazz of Ornette Coleman, John Coltrane, Cecil Taylor, and Albert Ayler. There were exceptions—Cornelius Cardew was a composer, Hugh Davies was classically trained and already an authority on international electronic music, Derek Bailey had played in dance bands and accompanied entertainers of all kinds—but John Stevens, Trevor Watts, Kenny Wheeler, Evan Parker, Paul Lytton, Jeff Clyne, Paul Rutherford, Keith Rowe, Lou Gare, Maggie Nicols, Eddie Prévost, Tony Oxley, and Lol Coxhill were all versed in jazz language, and to some degree they conceptualized their improvised music practice as both a development from and break with the predominantly African American jazz tradition.

During this period, between 1965 and the early 1970s, the rhythmic pulse and harmonic framework of jazz diminished in importance for such players. For some of them it was discarded entirely, which is one of the reasons why the music came to be known as free improvisation. Inevitably, there were connections with libertarian social trends, collectivist political ideas, and utopias of personal freedom, though they were less strong than in certain strands of rock, soul, folk, and even the avant-garde of composition. The deeper motivation of improvisation was driven by possibilities of liberation from structures and rules that were seen to be exhausted and excessively restrictive. Improvisers began to think of their music as textural, or as a nonvolitional (in the sense of giving the impression of stasis) interplay between the participants in a group. The collective sound was more important than hierarchies or displays of virtuosity (though virtuosity remained central to musical identity for the first generation of musicians, and some who followed) and so the technology used in this context was less dependent on established standards of uniformity, reliability, and flexibility expected from conventional musical instruments. Any artifact that could make a sound, transform sound, or resonate another sound could be incorporated into this relatively open framework.

There were precedents for this openness. The use of so-called exotic instruments in rock and jazz, whether the sitars and tabla of The Beatles and Miles Davis or the shofar, argol, and sarewa of Yusef Lateef, was a logical outcome of growing curiosity about other musical systems. The questions of whether listening itself was entrained by orthodoxies of Equal Temperament tuning, or whether existing musical instruments such as the piano or drum kit were machines engineered to generate particular musical forms, seemed at the heart of a problem: how could music making struggle free from its existing boundaries, its hierarchies, the constraints of its contexts, its spaces, and even the more conservative expectations of its audiences?

One pragmatic response to this interrogation of musical process was to amplify objects. Hugh Davies was one of the pioneers of this approach. After completing his degree in Oxford, Davies moved to Cologne to become an assistant to Karlheinz Stockhausen. In 1964, he performed in the ensemble that recorded Stockhausen's *Mikrophonie I*, a piece in which the surface activation of a large tam-tam is amplified with contact microphones. Stockhausen referred to this technique, developed for him by technician Jaap Spek, as a form of probing, "as a doctor auscultates a body with his stethoscope."[2] Informed by this experience and advised by Spek, Davies returned to England and developed his own low-tech, often humorous version of auscultation.

Although his inventions became visual artifacts, shown for example in the exhibition *New and Rediscovered Musical Instruments* (curated by Davies at the Scottish National Gallery of Modern Art, Edinburgh, 1975–1976) the instrument could also be perceived as an invisible, or immaterial, flow of information between object, performer, amplification device, and loudspeaker. This is not so different from Brian Eno's idea that the instrument was the recording studio and the score was the operational diagram of whichever system was used to produce the music. Many of the materials used by Davies as the basis of his live setup might ostensibly appear mundane as objects and inert as sound sources—an ordinary kitchen egg slicer (Figure 10.1), for example, or stretched

FIGURE 10.1
An ordinary kitchen egg slicer, once amplified and played by Davies, was transformed into a percussion instrument of great potential. From the Hugh Davies Collection at the Science Museum. Courtesy of the Science Museum (Inv: 2007-133). Images available from the Science & Society Picture Library.

springs—but once amplified and played by Davies, often with tools also taken from the bathroom cabinet, kitchen, and toolshed, they were transformed into percussion instruments of great potential.

Set in motion by early twentieth-century artists such as Marcel Duchamp, Man Ray, and Kurt Schwitters, the reenchantment of everyday objects formed an important part of the aesthetics and methodology of new instrument inventors. Surrealist objects such as Marcel Duchamp's *With Hidden Noise* (1936), or Oscar Dominguez's *Never* (1937), proposed an unheard music, fetishistic, secret and silent perhaps, but also liberating in the sense that they enter existence as artifacts through which imaginary music, the music of dreams, might be produced. Sound is occluded, reduced to nothing, yet amplified as an expression of the mythically resonant object. In the case of the Duchamp, a small object unknown to the artist was placed within a ball of twine encased between two plates of brass. Duchamp said, "It was a sort of secret and it makes a noise . . . Listen to it. I will never know whether it is a diamond or a coin."[3] As for *Never*, Dominguez assembled this "assisted ready-made" from a gramophone and a pair of mannequin legs. The female legs protrude, high-heeled shoes outward, from the bell of a Victrola horn; the other end of the horn is fashioned into a female hand which hovers over a breast-shaped turntable.[4] Like the body itself, or caves, pits, and dwellings, instruments are chambers, dark depths, entrances, and openings, sites of concealment and crucibles of origination.

Pablo Picasso's fascination with stringed instruments—the guitar, mandolin, and violin—as subjects for painting, sculpture, and assemblage might be considered as a premonition of the deconstruction of auditory artifacts. Picasso is said to have disliked music, so the question might be asked, why was he so interested in musical instruments? Perhaps this could be answered with the concept of space embodied within musical artifacts. One of the most economical ways of representing a contained space is to paint an acoustic guitar, itself a small, hidden room with a single entry point that allows sound to escape and project from the space within which the sound circulates and grows resonant and rich in overtones. The guitar or violin becomes, in a sense, a machine through which this interior space can be turned inside out and exist in many simultaneous versions of itself. This is indeed what happens when music takes place. The spectators watch an instrument and its activation, as if the instrument itself were the music, and yet the music is moving through and out of the instrument, joining itself to each individual listener and fusing them, along with all of the musicians, into the conglomerate that we term an audience. Music is an irruption, a cutting and filling, a dispersal and an absence. Picasso's *Guitar*, a construction of canvas, wood, rope, nails, and tacks on a painted panel, from spring 1926, confronts this invasive violence of music with its own violence: a menacing sound hole torn out of sacking, the nails protruding points out, and then a second image more like the soft imprint of the thought of a guitar, pressed into sand, though the tension of the guitar strap suggests a drawn bowstring, as if sound is an arrow that penetrates the heart.

Once the instrument was disassembled, turned inside out or mixed with other materials, then the normal functionality and economy of an instrument—its human scale, its versatility, the careful evolution of its specialization and the ease with which sound is produced by an able-bodied, practiced individual—could be discarded. This transformation of the economy of the artifact would inevitably disconnect the production of sound from music in its familiar forms (impossible to play Liszt, as we are accustomed to hear his music, on a prepared piano, or Vivaldi on a flute that can only play one pitch and its overtones). The activation of acoustic vibration requires resonance of some kind (in normal circumstances a chamber), but through amplification or by linking the object with some external chamber or surface, inert materials could be made to sound.

A 3D photograph, a length of elastic, or a coil of wire could be the starting point for a musical artifact. In the case of Max Eastley's *Aeolian Ground Harp*,[5] perhaps best described as a hypothetical or imagined instrument, the text proposed that a hole would be dug in the ground underneath a tree branch, then strings stretched from a crossbar attached to the branch down to a board covering the hole. The intention of the instrument, later refined and developed to great sophistication and effectiveness by Eastley, was a meeting point of nature and culture: wind would play across the strings to produce unearthly harmonies that were neither unadulterated meteorological or arboreal sounds, nor the intellectual decisions and physical manipulation of a human performer. Instruments could also be wearable, as with Anna (now Annea) Lockwood's *Sound-Hat* of hanging shells and sticks,[6] the details of which were published in Henri Chopin's *OU* magazine in 1970, and my own *Shell Hat* (1973),[7] a self-enclosing voice disguiser of shells and stones published as text and drawing in *New/Rediscovered Musical Instruments* (Figure 10.2).

In the latter case, I was inspired by voice disguisers, by Australian Aboriginal Kurdaitcha shoes—shoes of blood, hair, and emu feathers used in magical killing (Figure 10.3)—and the costume of the Yakut shaman, hung with metal representations of sacred creatures that would act as familiars for the shaman in his journey to spirit regions—in other words, specialist apparel with a magical rather than practical function. During the Victoria & Albert Museum's *Body Box* exhibition (1975), *Shell Hat* was "performed" as a vocal piece. Both the device and its performative aspect embodied my prioritization at that time of the symbolic function of musical instruments. In unpublished notes written as an addendum to *Shell Hat* I wrote the following: "By inference / a gesture > defetishization / acoustic technology / shell hat / musical instrument = muzzle / dog – weighs heavily on wearer / impedes song + hearing of song / sight / feeling of air currents + temperature changes on face + head / weight diverts concentration after a time—observers become intimidated by creature / becoming brave—touching / talking to / about it—response identical to responses observed in zoo situations."

I had become fascinated by voice disguisers in 1971 to 1972, during research in the BBC sound archives housed at Broadcasting House, London. Invited by BBC Radio Three archive producer Madeau Stewart to create a program, I collaged together

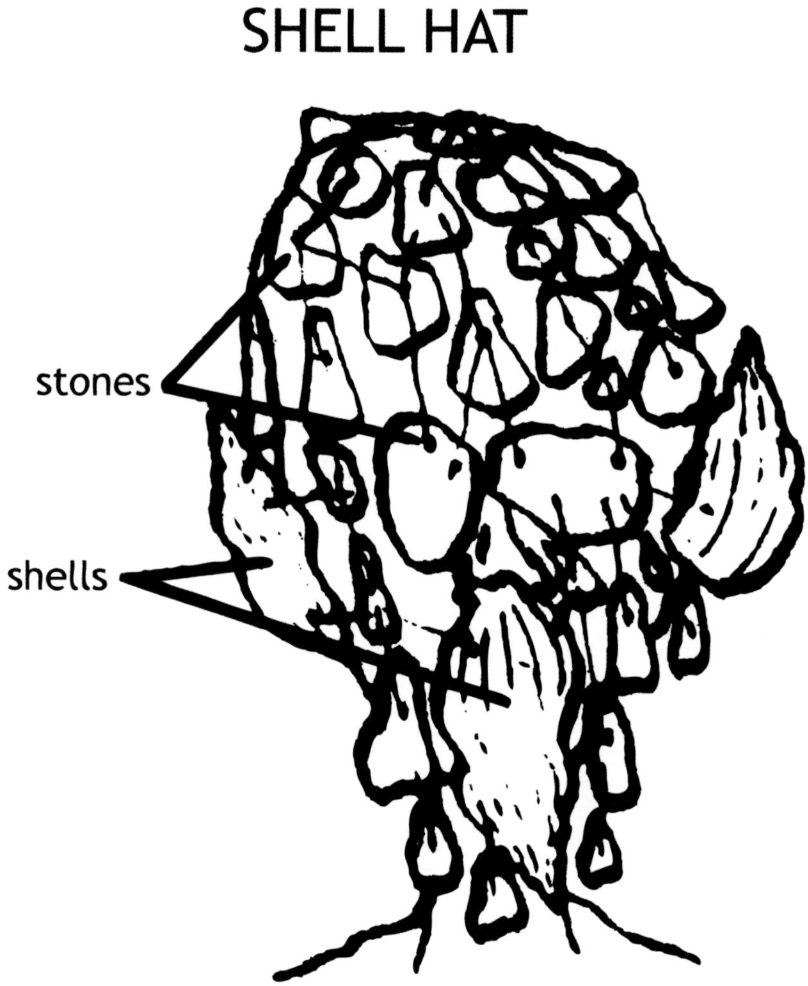

FIGURE 10.2
David Toop's *Shell Hat* (1973), a self-enclosing voice disguiser of shells and stones, drawing from *New/Rediscovered Musical Instruments* (1974). Courtesy of David Toop.

recordings of global musics and bioacoustic sounds. This was broadcast on BBC Radio Three as *Crossthreads* in May 1972. Voice disguisers were particularly interesting because of their transformative effect on the human voice and the affect of this transformation within ritual and ceremonial representations of supernatural forces and the extrahuman world. Recordings of voice disguisers were relatively rare. An exception was Hugo Zemp's *Masques Dan*,[8] released on the Ocora label. Zemp studied the music of the Dan people extensively. Made between 1965 and 1967 in an area bounded by Ivory Coast, Guinea, and Liberia, his recordings documented those masks that manifested a supernatural being through auditory means. In his notes to the record, Zemp makes the important point that many African masks could possess a dual function or be solely auditory:

FIGURE 10.3
Australian Aboriginal Kurdaitcha shoes, inspiration for *Shell Hat*, Wellcome Collection, Science Museum. Courtesy of the Science Museum (Inv: A266733). Images available from the Science & Society Picture Library.

> The appearance of African masks, above all the face mask, is well-known all over the world thanks to collections and books on African art; but their voices are much less familiar. The voice is, nevertheless, an essential part of the performance when the mask appears. True, mute masks exist; but there are others which are exclusively sound masks, i.e. they assume no visual disguise.[9]

The mask itself might produce a transformative effect on the wearer's voice and so become the spirit, or in the case of Guéwova, "Mask with a big voice," the voice of the spirit was played by two bullroarers swung simultaneously. Combinations of these voices could be complex in their inclusion of differing modalities of mask. Three circumcision masks were recorded by Zemp in Doualé in 1967: Wodoneugo, Baguédio, and Guéyomlo. The first of these was produced by a man singing and speaking with a guttural voice; the second by a man's voice singing into a hidden mirliton (among the Dan this might be a hornbill bone closed at one end with a thin, vibrating reed); the third by twenty young men rhythmically striking the ground with sticks as they sang in chorus. According to Zemp, "The voice of this mask literally shakes the ground."[10]

One of the most important influences on the contributors to *New and Rediscovered Musical Instruments* came from such global musics and musical instruments, most of which lay outside the mainstream of listening and commercial music production at that time. Paul Burwell's *5 Vellum Studded Drums*[11] were constructed from a mixture of industrial and synthetic materials—each drum fitted with two mirlitons made from rice-paper membranes. Burwell had heard recordings of African musicians playing drums and xylophones incorporating this principle and was excited by the way in which the sound of struck wood or skin was enhanced by the buzzing sound of a membrane (often the fragile casing of a spider's egg sac). In our duo of the early 1970s, Rain in the Face, there was a conscious effort to modify the timbre of existing instruments and to blur the distinction between acoustic and electronic masking. My own use of vintage distortion pedals for electric guitar such as the Rush Pep Box and Burwell's experimentation with acoustic mirlitons were attempts to construct an ambiguous space in which the nature of an instrument could transmigrate and become other than itself.

In her introduction to our book, Madeau Stewart wrote the following: "Where isolated tribes used insects as buzzers attached to or trapped in some casually cut length of local wood, so the new instrument makers—who are also composers—casually adapt a clock spring or an ex-army shell case as a sound maker. (Incidentally, a society surrounded by disposable or throwaway goods is not new.) Sometimes the wind is pressed to perform, and often the abundant sources of electricity are channelled to contribute to the creation of a composition. And audiences are mobile, free to chat and circulate as in the earliest days of the Prom."[12]

As Madeau Stewart was suggesting, the conditions that shaped the experience of sound work were expanding on all fronts and so all aspects of working with sound or experiencing sound were under interrogation. If our society had become so rigid and routine in its orthodoxies, then how could listening regain collective and individual intensity, a feeling of purpose and deeper meaning? Such questions, flawed as they might seem at a distance of nearly forty years, are nevertheless an indication of why more remote musical cultures seemed to offer intriguing alternatives. The search for new sounds was important in itself. Musicians had been searching for new sounds in all forms of music during the latter half of the twentieth century, whether the pedal steel guitar experiments of country music, the extended techniques of orchestral players and singers, the electronic distortions of rock, the innovative use of drum machines and synthesizers in soul and funk, or the overblowing of saxophones common to R&B and free jazz. "The Heteroglottal Clarinet," one of Evan Parker's contributions to *New and Rediscovered Musical Instruments*, was based on a modification of an illustration from K. G. Izikowitz's book, *Musical Instruments of the South American Indians* (also a wonderful resource for my own experiments in flute making). Parker fitted a grooved length of hardwood dowelling and a thin strip of bamboo within a large bamboo tube; the bamboo strip acted as a reed, with Parker's modification consisting of fishing lead folded over the exposed end of this reed. "Can be flicked rhythmically while playing to produce vibrato effect—like a musical bumble bee," he wrote (Figure 10.4).[13]

FIGURE 10.4

Wasp flute from *New/Rediscovered Musical Instruments* (1974). Here, enticed by jam, a wasp produces a musical drone. "After a short performance on the flute, the well-fed wasp is then carefully released." Courtesy of David Toop.

Many of the instruments invented by Parker and percussionist Paul Lytton could be heard in the early years of their improvising duo, both live and on record, though over time they were set aside, as if the timbral innovations they uncovered had been absorbed into more conventional tools—the soprano and tenor saxophone and the drum kit. Parker, in particular, was an avid listener to musics from South America, Tibet, Korea, Greece, Africa and so on; his singular approach to multiphonics, for example, was to some degree influenced when he was a young man by hearing performances by Greek and Indian musicians. The growth of interest in ethnomusicology during the 1960s was a stimulus to all of us as musicians but at the same time the discipline was affected by a growing feeling that remote musical cultures and their instruments could be living contributions to new soundwork, rather than simply lost traditions and inert objects. Musicians such as Richard Teitelbaum and Ivan Vandor, both members of the Rome-based improvising group Musica Elettronica Viva, were equally comfortable within ethnomusicology, electronic music, and improvisation.

Museum collections such as the Horniman Museum in London, the Pitt Rivers in Oxford, and the Haags Gemeentemuseum in The Hague were particularly valued for the opportunities they offered to see the arcane diversity of global musical instrument technology at close quarters, though these institutions were not so liberal as the Berlin Staatliche Museen under the direction of Dr. Artur Simon, where improvising musicians were allowed to take musical instruments from their cases and play them in performance (Evan Parker was, I believe, involved with this event in the 1970s).

For three of us whose work was published in *New and Rediscovered Musical Instruments*—Max Eastley, Paul Burwell, and me—along with Steve Beresford (an improvising musician better known at that time for using toys in his repertoire)—the culmination of this phase of instrument building came with *Whirled Music*, an improvised piece for whirling and spinning instruments. In 1977, Burwell concluded a performance at the London Musicians Collective by violently whirling heavy Chinese cymbals in wide circles and banging them on the floor. This produced spectacular Doppler effects along with a sense of clear and present danger. Max Eastley was inspired by the powerful visual and acoustic effect of this whirling and suggested a performance created entirely by whirling instruments. Although the initial discovery that led Burwell to spin Chinese cymbals in this way was the bullroarer, an eminently simple means to create the complex effect of supernatural presence, other spinning instruments used by the group were connected with more contemporary rituals. Large, wooden cog rattles were once used for air-raid warnings during wartime and later spun by football fans to encourage their teams with a burst of noise.

This returns us to the words of Hugo Zemp, who stressed that the auditory examples of masks recorded among the Dan people were a nexus point of both human and extra-human, as well as being a confluence of artifact and sound event. If one component of this formulation is subtracted, then it could be argued that the event as a whole ceases to exist. With the bullroarer, all that remains is a slender strip of wood and a length of string, stripped of its unsettling sounds, its threatening physicality and its symbolic power as an agent of cultural transformation. The same could be said of many of the instruments discussed in this chapter. The ostensibly disposable, humble artifacts amplified by Hugh Davies were not objects of conventional beauty. Despite this, or because of it, the potentiality of this way of working has been hugely influential, even in the present era when out-of-the-box applications for digital music composition are extensively available, easy to use, often free of charge, and capable of a high standard of production values. Nicolas Collins's book, *Handmade Electronic Music*, and recent recordings such as *Seven Vignettes* by Lee Patterson, with its improvised and edited sound pieces derived from the sounds of burning matches, the discharge of butane gas from discarded cigarette lighters, Andrews Liver Salts dissolving in water-filled wineglasses, burning hazelnuts, and the contact microphone recordings of heating pipes continue a now solidly established practice of transforming humble artifacts into a form of magic.

Notes

1. David Toop (ed.), *New/Rediscovered Musical Instruments* (London: Quartz/Mirliton, 1974).
2. Karlheinz Stockhausen, *Microphonie I and II/Telemusik*, Stockhausen 9 (CD), Germany, 1995, 19.
3. Marcel Duchamp, quoted in Jennifer Mundy (ed.), *Duchamp, Man Ray, Picabia* (London: Tate Publishing, 2008), 31.
4. Alyce Mahon, *Surrealism and the Politics of Eros: 1938–1968* (London: Thames & Hudson, 2005), 38.
5. Max Eastley in Toop, *New/Rediscovered*, 20.
6. Anne Lockwood, in *Henri Chopin's Revue OU*, Alga Marghen, O45U (4 x CD + book), Italy, 2002, no page number.
7. Toop, *New/Rediscovered*, 22.
8. Hugo Zemp, *Masques Dan: Côte d'Ivoire*, Disques Ocora, OCR 52 (LP), France, date unknown.
9. Ibid.
10. Ibid.
11. Paul Burwell in Toop, *New/Rediscovered*, 12.
12. Toop, *New/Rediscovered*, 2.
13. Ibid., 14.

Bibliography

Collins, Nicolas. *Handmade Electronic Music: The Art of Hardware Hacking*. New York: Routledge, 2009.

Izikowitz, Karl Gustav. *Musical Instruments of the South American Indians*. Yorkshire, UK: S.R. Publishers/The Galpin Society, 1970.

Lockwood, Anne. In *Henri Chopin's Revue OU*. Alga Marghen, O45U (4×CD + book), Italy, 2002.

Mahon, Alyce. *Surrealism and the Politics of Eros: 1938–1968*. London: Thames & Hudson, 2005.
Mundy, Jennifer, ed. *Duchamp, Man Ray, Picabia*. London: Tate Publishing, 2008.
Patterson, Lee. *Seven Vignettes*. Shadazz 07 (CD), UK, 2009.
Stockhausen, Karlheinz. *Microphony I and II/Telemusik*, Stockhausen 9 (CD), Germany, 1995.
Toop, David, ed. *New/Rediscovered Musical Instruments*. London: Quartz/Mirliton, 1974.
Zemp, Hugo. *Masques Dan: Côte d'Ivoire*, Disques Ocora, OCR 52 (LP), France, date unknown.

Artefacts Music Reading List

Albiez, Sean, and David Pattie, eds. *Kraftwerk: Music Non-Stop*. London: Continuum, 2011.

Alderman, John. *Sonic Boom. Napster, MP3, and the New Pioneers of Music*. Cambridge MA: Persus Publishing, 2001.

Auslander, Philip. *Liveness: Performance in a Mediatized Culture*, second, revised edition. London: Routledge, 2008.

Ayers, Michael D. *Cybersounds: Essays on Virtual Music Cultures*. New York: Peter Lang, 2006.

Bijsterveld, Karin, and José van Dijck, eds. *Sound Souvenirs: Audio Technologies, Memory and Cultural Practices*. Amsterdam: Amsterdam University Press, 2009.

Braun, Hans-Joachim, ed. *Music and Technology in the Twentieth Century*. Baltimore, MD: The Johns Hopkins University Press, 2002.

Brend, Mark. *Strange Sounds: Offbeat Instruments and Sonic Experiments in Pop*. San Francisco: Backbeat Books, 2005.

Brøvig-Hanssen, Ragnhild. "Opaque Mediation: The Cut-and-Paste Groove in DJ Food's 'Break.'" In *Musical Rhythm in the Age of Digital Reproduction*, pp. 159–75, ed. A. Danielsen. Farnham, UK: Ashgate, 2010.

Bull, Michael. *Sound Moves: iPod Culture and Urban Experience*. Oxford: Routledge, 2008.

———. *Sounding Out the City: Personal Stereos and the Management of Everyday Life*. Oxford, UK: Berg, 2000.

Bull, Michael, and Les Back, eds. *The Auditory Culture Reader*. Oxford: Berg, 2000.

Chadabe, Joel. *Electric Sound: The Past and Promise of Electronic Music*. Upper Saddle River, NJ: Prentice Hall, 1997.

Chanan, Michael. *Repeated Takes: A Short History of Recording and Its Effects on Music*. London: Verso, 1995.

Chew, Victor K. *Talking Machines, 1877–1914: Some Aspects of the Early History of the Gramophone*, second edition. London: H.M.S.O., Science Museum, 1982.

Cox, Christoph, and Daniel Warner, eds. *Audio Culture: Readings in Modern Music*. New York: Continuum, 2004.

Daniel, Eric D., C. Denis Mee, and Mark H. Clark, eds. *Magnetic Recording: The First 100 Years*. New York: IEEE Press, 1999.

Danielsen, Anne, ed. *Musical Rhythm in the Age of Digital Reproduction*. Farnham: Ashgate, 2010.

Davies, Hugh. "Gentle Fire: An Early Approach to Live Electronic Music." *Leonardo Music Journal* 11 (2001): 53–60.

Day, Timothy. *A Century of Recorded Music: Listening to Musical History*. New Haven, CT: Yale University Press, 2000.

DeNora, Tia. *Music in Everyday Life*. Cambridge, UK: Cambridge University Press, 2000.

d'Escriván, Julio. "To Sing the Body Electric: Instruments and Effort in the Performance of Electronic Music." *Contemporary Music Review* 25.1/2 (2006): 183–91.

Dolan, Brian. *Inventing Entertainment: The Player Piano and the Origins of an American Musical Industry*. Lanham, MD: Rowman & Littlefield, 2009.

Donhauser, Peter. *Elektrische Klangmaschinen: Die Pionierzeit in Deutschland und Österreich*. Vienna: Böhlau, 2007.

Doyle, Peter. *Echo and Reverb: Fabricating Space in Popular Music Recording, 1900–1960*. Middletown, CT: Wesleyan University Press, 2005.

Eisenberg, Evan. *The Recording Angel: Music, Records and Culture from Aristotle to Zappa*. New Haven, CT: Yale University Press, 2005.

Emmerson, Simon. *Living Electronic Music*. Farnham, UK: Ashgate, 2007.

Fischer-Defoy, Christine. *Kunst, Macht, Politik: Die Nazifizierung der Kunst- und Musikhochschulen in Berlin*. Berlin: Elefanten Press, 1988.

Forrest, Peter. *The A–Z of Analogue Synthesisers*. Exeter, UK: Susurreal, 1998, 2003.

Frith, Simon. "Art versus Technology: The Strange Case of Popular Music." *Media, Culture & Society* 8 (1986): 263–79.

Frith, Simon, and Andrew Goodwin, eds. *On Record: Rock, Pop and the Written Word*. London: Routledge, 1990.

Glinsky, Albert. *Theremin: Ether Music and Espionage*. Champaign: University of Illinois Press, 2005.

Gracyk, Theodore. *Rhythm and Noise: An Aesthetics of Rock*. Durham, NC: Duke University Press, 1996.

Greene, Paul D., and Thomas Porcello, eds. *Wired for Sound: Engineering and Technologies in Sonic Cultures*. Middletown, CT: Wesleyan University Press, 2005.

Hankins, Thomas. L., and Robert. J. Silverman. *Instruments and the Imagination*. Princeton, NJ: Princeton University Press, 1995.

Hegarty, Paul. *Noise/Music: A History*. London: Continuum, 2007.

Holmes, Thom. *Electronic and Experimental Music*. New York: Routledge, 2002.

Hoover, Cynthia A. *Music Machines—American Style. A Catalog of the Exhibition*. Washington DC: Smithsonian Institution Press, 1971.

IMA Institut für Medienarchäologie. *Zauberhafte Klangmaschinen. Von der Sprechmaschine bis zur Soundkarte*. Hainburg: Schott, 2008.

Jüttemann, Herbert. *Phonographen und Grammophone*. Braunschweig, Germany: Klinkhardt & Biermann, 1979.

Katz, Mark. *Capturing Sound: How Technology Has Changed Music*. Berkeley: University of California Press, 2010.

Kraft, James P. *Stage to Studio: Musicians and the Sound Revolution, 1890–1950*. Studies in Industry and Society 9. Baltimore, MD: The Johns Hopkins University Press, 1996.

Kvifte, Tellef. "Digital Sampling and Analogue Aesthetics." In *Aesthetics At Work*, ed. Arne Melberg, 105–28. Oslo: Unipub, 2007.

LaBelle, Brandon. *Acoustic Territories: Sound Culture and Everyday Life.* London: Continuum, 2010.

Laurendeau, Jean. *Maurice Martenot: Luthier de L'Électronique.* Croissy-Beaubourg, France: Dervy-Livres, 1990.

Lysloff, René T. A., and Leslie C. Gay, eds. *Music and Technoculture.* Middletown, CT: Wesleyan Press, 2003.

Manning, Peter D. *Electronic and Computer Music: Revised and Expanded Edition.* New York: Oxford University Press, 2004.

Millard, Andre. *America on Record: A History of Recorded Sound.* Cambridge, UK: Cambridge University Press, 1995.

Milner, Greg. *Perfecting Sound Forever: The Story of Recorded Music.* London: Granta Books, 2009.

Morton, David. *Off the Record: The Technology and Culture of Sound Recording in America.* New Brunswick, NJ: Rutgers University Press, 2000.

———. *Sound Recording: The Life Story of a Technology.* Baltimore, MD: The Johns Hopkins University Press, 2004.

Niebur, Louis. *Special Sound: The Creation and Legacy of the BBC Radiophonic Workshop.* Oxford, UK: Oxford University Press, 2010.

Nyman, Michael. *Experimental Music: Cage and Beyond.* Cambridge, UK: Cambridge University Press, 1999.

Ochse, Orpha. *The History of the Organ in the United States.* Bloomington: Indiana University Press, 1988.

Philip, Robert. *Performing Music in the Age of Recording.* New Haven, CT: Yale University Press, 2004.

Pinch, Trevor J., and Karin Bijsterveld. "'Should One Applaud?' Breaches and Boundaries in the Reception of New Technology in Music." *Technology and Culture* 44(3) (2003): 536–59.

———. "Sound Studies: New Technologies and Music." *Social Studies of Science* 34(5) (2004): 635–48.

———. eds. *The Oxford Handbook of Sound Studies.* New York: Oxford University Press, 2011.

Pinch, Trevor, and Frank Trocco. *Analog Days: The Invention and Impact of the Moog Synthesizer.* Cambridge, MA: Harvard University Press, 2002.

Prieberg, Fred K. *Musik im NS-Staat.* Frankfurt am Main: Fischer Taschenbuch Verlag, 1982.

Rabinovici, Alison. "Augustus Stroh's Phonographic Violin. A Journey: Victorian London, Australia, Transylvania." *Galpin Society Journal* 58 (May 2005): 100–23.

Read, Oliver, and Walter L. Welch. *From Tin Foil to Stereo: Evolution of the Phonograph.* Indianapolis, IN: Howard W. Sams, 1959.

Roell, Craig. *The Piano in America, 1890–1940.* Chapel Hill: University of North Carolina Press, 1989.

Sadie, Stanley, ed. *The New Grove Dictionary of Musical Instruments.* London: Macmillan Press Ltd., 1984.

Schafer, R. Murray. *The New Soundscape: A Handbook for the Modern Music Teacher.* Toronto: Berandol, 1969.

———. *The Soundscape: Our Sonic Environment and the Tuning of the World.* Rochester, VT: Destiny Books, 1994.

Schrader, Barry. *Introduction to Electro-Acoustic Music*. Englewood Cliffs, NJ: Prentice-Hall Inc., 1982.

Sterne, Jonathan. *The Audible Past: Cultural Origins of Sound Reproduction*. Durham, NC: Duke University Press, 2003.

Taylor, Timothy D. *Strange Sounds: Music, Technology and Culture*. New York: Routledge, 2001.

Théberge, Paul. *Any Sound You Can Imagine: Making Music/Consuming Technology*. Hanover, NH: Wesleyan University Press, 1997.

Thompson, Emily. *The Soundscape of Modernity*. Cambridge, MA: MIT Press, 2002.

Toop, David. *Ocean of Sound: Aether Talk, Ambient Sound and Imaginary Worlds*. London: Serpent's Tail, 1995.

———. *Haunted Weather: Music, Silence and Memory*. London: Serpent's Tail, 2004.

———. *Sinister Resonance: The Mediumship of the Listener*. London: Continuum, 2010.

Toynbee, Jason. *Making Popular Music: Musicians, Creativity and Institutions*. London: Arnold, 2000.

Ungeheuer, Elena, ed. *Elektroakustische Musik. Handbuch der Musik im 20. Jahrhundert. Band 5*. Laaber, Germany: Laaber-Verlag, 2002.

———. "Imitative Instrumente und innovative Maschinen? In IMA Institut für Medienarchäologie, *Zauberhafte Klangmaschinen. Von der Sprechmaschine bis zur Soundkarte*. Hainburg, Germany: Schott, 2008.

Vail, Mark. *The Hammond Organ: Beauty in the B*. San Francisco: Backbeat Books, 2002.

Veal, Michael E. *Soundscapes and Shattered Songs in Jamaican Reggae*. Middletown, CT: Wesleyan University Press, 2007.

Weidenaar, Reynold H. *The Telharmonium: A History of the First Music Synthesizer, 1893–1918*. Dissertation, New York University, 1989.

Weium, Frode. "Ingenieurmusik. Die Begegnung mit elektrischen Musikinstrumenten in Norwegen in den 20er- und 30er-Jahren." *Blätter für Technikgeschichte* 69/70 (2008): 135–54.

Young, Gayle. *The Sackbut Blues: Hugh Le Caine, Pioneer in Electronic Music*. Ottawa: National Museum of Science and Technology, 1989.

Young, Rob. *Undercurrents: The Hidden Wiring of Modern Music*. London: Continuum, 2002.

Index

About the Contributors

Sarah Angliss is a composer, performer, and maker whose artistic work is inspired by the history of technology, including her own archival discoveries. She is the founding member of the human, theremin, and robot band Spacedog. In 2003, Sarah instigated and led *Infrasonic*, an exploration of the curious psychological effects of extreme bass noise (Purcell Room, London). This led to her collaboration with Punchdrunk Theatre Company to create an unusual special effect for their immersive documentary *It Felt Like a Kiss* (Manchester International Theatre Festival, 2009). Sarah's live performance exploring Lancashire clog dancing as early music (with Dr. Caroline Radcliffe) was awarded a Quake Festival Award (2006) and was featured in her TEDx talk *Loving the Machine* in 2011. In July 2011, Sarah wrote and presented the BBC Radio 4 documentary *The Bird Fancyer's Delight*, which explored the use of trained birds as primordial feathered sound recorders. She is currently collaborating with music labels Ghostbox and Gecophonic and composing new music from found sounds for *The Effect,* a new play by Lucy Prebble at London's National Theatre (director Rupert Goold). Contact Sarah Angliss at hello@ sarahangliss.com.

Tim Boon is head of research and public history at the Science Museum, London. His exhibitions include *Health Matters* (1994), *Making the Modern World* (2000), and *Oramics to Electronica* (2011). He was curator of the Icebreaker concert performances of Brian Eno's *Apollo* in 2009 to mark the fortieth anniversary of the moon landings. He has a long-standing personal interest in music—he even built his own synthesizer in the 1980s—and has recently begun to develop academic interests in this area. His first book, *Films of Fact: A History of Science in Documentary Films and Television* (Wallflower Press) was published in 2008. He is currently running the Science Museum's Public History project, which is experimenting with participatory exhibitions with the aim of developing more effective displays on the history of science and technology. Contact Tim Boon at the Science Museum, Exhibition Road, London SW7 2DD, tim.boon @sciencemuseum.ac.uk.

Ragnhild Brøvig-Hanssen is a PhD student at the Department of Musicology, University of Oslo. Her PhD project, *Opaque Mediation*, concerns aesthetic changes in popular music after the digitalization of technology. She holds an MA in musicology from the University of Oslo and received a master's scholarship in 2005 from the research project *Rhythm in the Age of Digital Reproduction*. She has published articles in Cambridge's journal *Popular Music* and the anthology *Musical Rhythm in the Age of Digital Reproduction*, ed. Anne Danielsen (Farnham, Surrey: Ashgate, 2010). In addition to her academic engagements, she is a producer, composer, and performer for the band Linen Cox. She has also taught music production at the University of Oslo. In 2008, she contributed in the planning of the exhibition *Music Machines* (2009) at the Norwegian Museum of Science and Technology. Contact Ragnhild Brøvig-Hanssen at the Department of Musicology, University of Oslo, P.O. Box 1017, Blindern, 0315 Oslo, Norway, ragnhild.brovig-hanssen@imv.uio.no.

Peter Donhauser is a physicist and musician who has worked since 1989 at the Vienna Museum of Technology, where he was director from 1997 to 2000, reopening the renovated museum. Since then he has served as head curator for scientific, mathematical, medical, geodesy, and musical instruments. Since the winter semester of 2011, he is also an assistant professor at Vienna University. His permanent exhibitions are *Nature and Knowledge* and *Musical Instruments*. Temporary exhibitions include *Playing with Technology*, *Power of Music*, and *Enchanting Sound Machines*, among others. In 2004, a research project about early electronic musical instruments led him to a deeper engagement in the topic. Publications include: *Restaurieren, Renovieren, Rekonstruiren* (Vienna, 1995), *Elektrische Klangmaschinen* (Vienna, 2007), *Streicher-100 Jahre Klavierbau in Wien* (forthcoming, 2012) and numerous articles in proceedings, catalogues, and more. Contact Peter Donhauser at Vienna Museum of Technology, Mariahilfer Strasse 212, 1140 Vienna, Austria, peter.donhauser@tmw.at.

Brian Eno is an artist and musician who has worked in the media of recorded sound and installation art. His records include such genre-creating works as *Discreet Music* (1975), *My Life in the Bush of Ghosts* (with David Byrne, 1981), *On Land* (1982), and *Drums between the Bells* (with Rick Holland, 2011). Among his installations are *77 Million Paintings* (2006) and *Music for White Cube* (1997). He has consistently smuggled ideas from the avant-garde into the mainstream, challenging both. Constantly in demand for opinions and insights, Brian may well have had as much of an influence on how people listen to, and think about, music as he has had on what they actually hear, despite his production work with some of the most successful pop music acts of the last three decades.

Mick Grierson specializes in applied real-time audiovisual interaction and cognition research. He has fifteen years of experience creating motion graphics, music, sound, and interactive installations for the arts, games, and entertainment industries, most recently as the consultant and software designer for Christian Marclay's award-winning installation *The Clock*. He is currently principal investigator on a three-year fellowship developing a range of interactive audiovisual and brain-computer interface software for platforms including consoles, iPhones, and iPads. Since 2006 he has been developing an archive to preserve and celebrate the work of Daphne Blake Oram. He is director of the Daphne Oram Collection at Goldsmiths College and a founder member of the Daphne Oram Trust, a charity devoted to supporting experimental and outsider research in the technology-led arts. Contact Mick Grierson at Goldsmiths, University of London, New Cross, London, SE14 6NW, UK, m.grierson@gold.ac.uk.

Aleksander Kolkowski is a composer, violinist, and researcher born and based in London. As a professional musician, he has performed worldwide and recorded for numerous labels with various groups and as a soloist. Since 1999, he has explored the potential of early sound recording and reproduction technology, using mechanical instruments, gramophones, and wax cylinder phonographs, to make contemporary mechanical-acoustic music. This work has been performed and broadcast internationally. His works include *Mechanical Landscape with Bird* (2004) for Stroh instruments, phonographs, serinette, and live singing canaries and *Horatio Oratorio* (2008), combining historic sound recordings with computer-generated music. He also has one of the largest private collections of mechanically amplified, horned string instruments in existence. Aleks is the London Science Museum's sound artist-in-residence during 2012. In June 2012, he resurrected the museum's 1905 Auxetophone for a concert, combining it with a ten-piece ensemble, pianola, and voice. Contact Aleksander Kolkowski at aleks@recording-angels.info.

Tellef Kvifte is full professor at the Department of Musicology at the University of Oslo. His research interests span from Norwegian hardanger fiddle music via theory of rhythm to theoretical organology and music technology, and he has published internationally in all these areas. His most recent research concerns perspectives on the codevelopment of music, music technology, notation, and concepts of sounds. Kvifte occasionally also appears on the professional World Music scene as a musician, and he performs tin whistles, hardanger fiddles, saxophones, laptops, and a variety of other instruments with confidence. He worked professionally as a television producer before taking up his academic career, and he is still a noted record producer. Contact Tellef Kvifte at the Department of Musicology, the Norwegian Collection of Folk Music, University of Oslo, P.O. Box 1017, Blindern, 0315 Oslo, Norway, tellef.kvifte@imv.uio.no.

Katy Price is lecturer in modern and contemporary literature at Queen Mary, University of London. Her first book, *Loving Faster Than Light: Romance and Readers in Einstein's Universe* is published in 2012. She has written and performed text works for performance with and without computers, including "Under the Yoke" (with Tom Hall) at the Berlin Supercollider Symposium, 2010, and "Bookmachine" at Kettle's Yard, Cambridge, 2010.

Contact Dr. Price at katyprice.wordpress.com, School of English and Drama, Queen Mary, University of London, Mile End Road, London E1 4NS, UK, k.price@qmul.ac.uk.

Alison Rabinovici is enrolled in a PhD (musicology) program at the University of Melbourne. Her recently completed master's thesis, *A History of Horned Strings: Organology and Early Sound Recording 1899–1945*, gained her an APA scholarship, and also the prestigious Helen Macpherson Smith Scholarship, an award made annually by the University of Melbourne to an outstanding woman graduate in the humanities. During her master's candidature she was also awarded the Percy Jones Award for Musicology. Her interest in the connections between museum artifacts and historical musicology led to publications in *Context: Journal of Music Research* (2005) and *The University of Melbourne Collections Journal* (2008). Her interest in Stroh violins and horned strings generally is long-standing, and her first publication, "Augustus Stroh's Phonographic Violin. A Journey: Victorian London, Australia, Transylvania," was published in *The Galpin Society Journal* in 2005. Rabinovici's interests include the areas of organology, nineteenth-century British and Australian performance and reception, and the history of recorded sound. Contact Alison Rabinovici at the University of Melbourne, Australia, alison_rabinovici@fastmail.fm.

David Toop is a musician, writer, and sound curator. His acclaimed books include *Rap Attack* (1984), *Ocean of Sound* (1995), *Haunted Weather* (2004), and *Sinister Resonance: The Mediumship of the Listener* (2010). His writing has also appeared in *The Wire*, *Bookforum*, and the *New York Times*. His discography spans nearly four decades. His first record, a collaboration with the sound sculptor Max Eastley titled *New and Rediscovered Musical Instruments*, was released in 1975 on Brian Eno's Obscure label. He is a senior research fellow at the London College of Communication and visiting professor at Leeds College of Music. Contact David Toop at London College of Communication, davidtoop@blueyonder.co.uk.

Frode Weium is a curator at the Norwegian Museum of Science and Technology in Oslo and PhD candidate at the Department of History and Classical Studies, Norwegian University of Science and Technology (NTNU) in Trondheim. His permanent exhibitions include *A Sound Mind in a Sound Body* (2003), *Music Machines* (2009), and *Instrument: Stories of Science* (2011). He has also contributed to temporary exhibitions on music and acoustics. He was editor of the museum book series *Volund* from 2000 to 2003. Over the last few years, he has written several articles on music technology, focusing on the history of early electronic musical instruments (such as the Theremin, the Ondes Martenot, and the Trautonium) and investigating broader questions about the relationship between technology and culture. His PhD project concerns the introduction and reception of electronic musical instruments in Norway. In 2007 he organized the twelfth Artefacts meeting in Oslo. Contact Frode Weium at the Norwegian Museum of Science and Technology, Kjelsåsveien 143, 0491 Oslo, Norway, frode.weium@tekniskmuseum.no.

Sean Williams is a Leverhulme Trust Research Fellow at the University of Edinburgh. He is an active performer, using live electronics as the founding member of Intuitive Music Edinburgh and the Monosynth Orchestra, playing original compositions, improvisations, and existing pieces by Stockhausen, Wolff, Subotnik, Ono, and others. His performance practice is directly connected with his research into the technical aspects of early electronic music. He designs and builds electronic instruments for performance and also enjoys playing records whenever and wherever the opportunity arises—at home, in performance, in seminars, in the Nevada desert—and has produced and presented a weekly radio show called *Voice on Record* for Resonance FM. His sound art pieces include installations for Lyrebird at the Threshold art space (Perth, 2007), Dialogues Festival (Edinburgh, 2009), University of Abertay (Dundee, 2010), and ZKM (Karlsruhe, 2011). Contact Sean Williams at University of Edinburgh, Reid School of Music, 12 Nicolson Square, Edinburgh, EH8 9DF, UK, sean@sbkw.net.